S0-BLM-080

SILICON and its Compounds

B. MAZUMDER

Science Publishers, Inc.
Enfield (NH), USA Plymouth, UK

CIP data will be provided on request.

SCIENCE PUBLISHERS, INC.
Post Office Box 699
Enfield, New Hampshire 03748
United States of America

Internet site: *http://www.scipub.net*

sales@scipub.net (marketing department)
editor@scipub.net (editorial department)
info@scipub.net (for all other enquiries)

ISBN 1-57808-154-8

Published by Science Publishers, Inc., Enfield, NH, USA
Printed in India

Dedicated to
my late mother, Mrs. S. Mazumder and
my daughter Indrani Mazumder

Preface

Silicon metal and its compounds have fascinated scientists and technologists on account of their extraordinary properties and enthused them to search for still newer compounds of silicon. While this process of uncovering compounds has been an uphill task for scientists, technologists and engineers have had to face the even more daunting task of translating laboratory results to actual plant design for commercial production.

The continued quest for finding newer techniques for production of high purity silicon and its compounds have produced bounteous results over the years. It now seems that while chances of discovering more techno-economically feasible routes to production of ultra pure silicon has almost reached saturation point, the search for newer silicon-compounds is still opening up vistas for both newer and innovative synthesis techniques and production of ever increasing number of exotic compounds like liquid crystals to fullerene-silicon compounds. The literature in this field is highly scattered and difficult to find in one place. An attempt has been made in this concise presentation to bring as much information as possible under one cover.

Most of the literature has been collected over last two decades. While older well-established results have been reviewed in short, more details have been provided for the references that appeared in last five years or so. The book will serve as a reference source for research workers in this area and as an important supplement to graduate text books in materials science, chemistry and chemical engineering as well.

B. Mazumder

Contents

CHAPTER 1

Introduction

1.1 SILICON AND ITS PHYSICAL PROPERTIES

One of earths most abundant resource element is silicon. Earth's crust contains about 27.6% silicon, whereas concentration of aluminium is 8%, iron 5%, calcium 3.6%, sodium 2.8%, potassium 2.6%, hydrogen 0.14%, and oxygen 46.4%. Yet it is difficult to obtain it in pure form because of its high affinity towards oxygen and other electronegative atoms. Since it occupies middle position in the group of elements of the periodic table (between metals and non-metals) it exhibits some extraordinary property which are of great value for industrial exploitation. Thus to obtain it in purest form has thrown challenge to engineers and scientists in the last few decades, while to synthesize its organometallic compounds in an economical way has become a matter of intense research in recent times.

Silicon crystallizes in tetrahedral structure like diamond, and has electrical resistance of about 400 k–ohm at room temperature. The resistivity decreases with temperature and increasing concentration of electrically active elements like boron, aluminium, galium, indium, thalium, phosphorus, arsenic and antimony. Presence of even a minute quantity of these elements in pure silicon, its room temperature resistivity reduces by 9 orders of magnitude. Except for very thin film, silicon is impermeable to visible light but is highly permeable to infrared light. The IR permeability decreases with decreasing electrical resistivity. Density of silicon at 300 K is 2.329 gs/cc, and volume increase during phase transformation from liquid to solid is +9.1%. Its thermal expansivity is 2.6×10^{-6}/K at 300 K and thermal conductivity at this temperature is 1.5 watt/cm/K. Bulk silicon is very much resistant to acid (including hydrofluoric and nitric acid), but not to a mixture of hydrofluoric and nitric acid. The reason being that, when silicon is treated with the above acid mixture, hydrofluoric acid first etches away the silica on its surface and makes it vulnerable to attack by nitric acid then. Other physical properties of silicon are shown in Table 1. Chemically it reacts with hydrochloric acid in the presence of a catalyst to form two useful products—silicon tetrachloride and trichloro-silane at about 570 K, and with methyl-chloride to form alkyl-chlorosilane.

Table 1. Physical properties of pure silicon crystals.

Structure	FCC (two displaced interpenetrating lattice) at 150,000 atmosphere FCC changes to BCC lattice.
Atomic density (atom/cc)	5×10^{22}
Melting point (°C)	1410
Boiling point (°C)	2355
Density (g/cc at 25°C)	2.329 (liquid silicon is denser 2.55 g/cc, because solid Si has open structure whereas liquid Si is typically metallic liquid).
Critical temperature (°C)	4886
Critical pressure (MPa)	53.6
Electronegativity	1.8 (hydrogen = 2.1 and carbon = 2.5)
Hardness (mhos scale)	6.5
Heat of fusion (kJ/g)	16 (surface-tension of liquid Si is about 720 mNm^{-1})
Band gap (E_g at 25°C) eV	1.12
Dielectric constant	11.8
Resistivity (at room temp.)	1000 ohm.cm (when impurity level is 10^{13}/cc)

Silicon atom itself, on the other hand, occurs in nature in three isotopic forms and their natural abundances are—^{28}Si (92.23%), ^{29}Si (4.67%), and ^{30}Si (3.10%). The ^{29}Si has a nuclear spin of +½ and has potential for nuclear magnetic resonance studies. Enriched (95%) ^{29}Si isotope costs about $ 26000/g while ^{30}Si isotope costs[1] $46000/g. Thus these isotopes have become of vital importance for science and medicine[2]. High purity silicon on the other hand possesses photo-electric property in ordinary sunlight and holds great promise for our future non-polluting and non-exhaustible power supply. This special property of silicon occurs because of its electronic structure and band-gap vis-a-vis energy available in sunlight. At 1991 price level, cost of metallurgical or crude variety of silicon was $ 1-3/kg (depending on purity), while that of photovoltaic or semiconductor grade was $ 50-70/kg. High purity silicon has impurity level of the order of ppm to ppb, and is available in monocrystalline (or single crystal), polycrystalline, and in amorphous variety (also written as Si : H). Before we go into production technology of silicon, let us first see the reason for the demand for such high level of purity for photovoltaic and other uses.

1.2 SILICON AS A PHOTOVOLTAIC MATERIAL

Figure 1 shows the absorption curve for various semiconductor elements. There are two distinctly different varieties of semiconductors, from light absorption characteristics point of view–one group in which absorption constant α varies very rapidly from very low value (at $h\alpha = E_g$) to values in excess of 10^4 cm^{-1} (e.g. GaAs and CdTe fall in this category), and the second group like Si and GaP whose α rises more gradually. The rapidly rising absorption curves are characteristics of 'direct band gap material', whereas the slowly rising curves are characteristics of 'indirect band gap materials'. Because of their rapidly rising absorption curve, direct gap semiconductors are better suited for photovoltaic cells, as the thickness of material required to absorb all photons in excess of E_g is smaller in direct band gap elements than indirect band gap materials; consequently less material will be needed for solar cell manufacture. Whereas 1 µm of GaAs would absorb about 80% of the maximum number of Air Mass O photons absorbable in this material, about 10 µm of Si would be needed to absorb 80% of the number absorbable in silicon. Moreover, because of the non-linear relationship among absorption coefficient, wavelength, and minimum ionization energy (1.12 eV for silicon), the silicon solar cells do not respond equally to all wavelengths of light. Thus infrared photons having a wavelength larger than 1.1 µm cannot release a hole-electron pair, and photons in the visible and ultraviolet region of the solar spectrum having excess energy merely generates heat in Si.

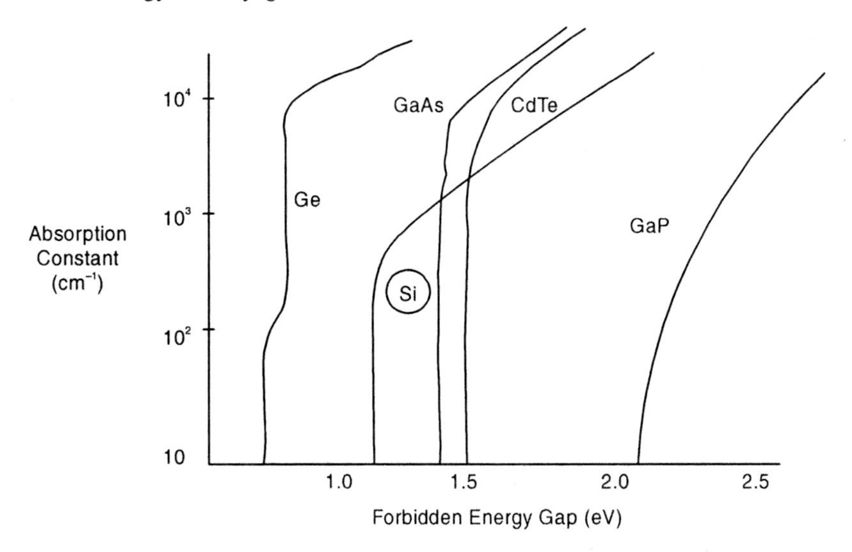

Fig. 1. Absorption curve of various semiconductors.

It may be noted here that, the spectrum and intensity (1 kW/m^2) of sunlight when the sun is directly overhead is referred to as 'Air-Mass-1' (AM1) sunlight. 'Air-Mass-0' (AM0) occurs in space sunlight in the vicinity of the earth (sunlight intensity 1.39 kW/m^2) and above the earth's atmosphere. As mentioned above, since silicon solar cells are sensitive to particular wavelengths in the spectrum of light, care should be taken in interpreting results of terrestrial and artificial light tests, and extrapolating results from such tests to space conditions. Our atmosphere absorbs and depletes sunlight of the photons in the 'blue' end of the spectrum. As a result stability of solar cells with extra bombardment of rays in extraterrestrial site, calls for use of best quality CZ-silicon cells (since efficiency of output per square metre is also an important parameter here), and float-zone (FZ) variety is not used for extraterrestrial applications. Market demand shows, CZ-silicon constitutes 85% of commercial output and FZ-silicon constitutes the rest 15%, in single crystal market. Since response of silicon solar cells is weaker in the blue region of the spectrum than red, consequently conversion efficiency for AM1 can be higher than for AM0 spectrum.

Commercially available silicon solar cells show conversion efficiency of around 18%, whereas some lab modules report it as high as 24%. CZ-type silicon crystals have higher resistivity 7-13 ohm.cm compared to 0.5-5 ohm.cm for terresones, as the cells should resist degradation by cosmic radiations. Polycrystalline silicons has a lower efficiency (around 10%), and amorphous silicon in the range 5-6%. But even with lower efficiency, amorphous silicon competes well over other varieties of silicon because of its edge over cost in most commercial applications.

Since photovoltaic is directly proportional to the incident light to the photocell, concentration of solar energy enhances photovoltaic effect. Thus attention has been paid to design apparatuses for concentrating solar energy. In this effort, besides mirror systems, special light concentrators have been improvised especially in space programme, through inflatable silicon filled fresnel lenses[3,4] and metallic reflectors[5]. But increasing solar radiation from normal 0.1 to 2.5 W/cm^2 causes temperature of solar cells to soar up from 30 to 220°C with consequent increase in short-circuit photocurrent, and decrease in the maximum power output, open-circuit photocurrent, efficiency of the cell, and fill-factor[6]. It has been found that, increase in photocurrent is caused by the change of light absorption with increase in temperature. However, decrease of photovoltage and fill-factor is caused by change in charge carrier concentration and narrowing the width of prohibited crystal zone. As a result, inverse current of p-n transition increases. Decrease in maximum power and efficiency coefficient is related to the decrease in open-circuit voltage and fill-factor with increasing temperature.

Quantitatively, when a silicon solar cell is exposed to AM0 sunlight, 24% of the energy in the incident beam is lost because the photon energy is too low to produce ionization in the silicon. In addition, 32.5% of the incident energy is lost because each photon having energy in excess of the band gap, can generate only one hole-electron pair; the excess energy is degraded into thermal energy. Out of the balance 43.5% energy, how much is delivered to the load depends upon collection efficiency Q, cell geometry (e.g. square-cells give more complete area coverage than round cells), reflection losses (which could be as high as 10%). After taking into account all these losses, a potential theoretical upper limit of photovoltaic efficiency for silicon has been assigned to 25%. Efficiency loss due to quality of cell, can be improved by decreasing resistivity of the base wafer, from which the cell is fabricated, from a value of 10 ohm.cm to about 0.01 ohm.cm, while retaining post-fabrication minority carrier life time of about 1, and by reducing surface losses so that blue photons can be utilized more efficiently. Some of these improvements have already been realized through fabrication of so called 'Violet Cells', where substantial increase in blue response is achieved by altering the geometry of fabrication procedure of cells.

At present 95% of all electronic devices are made from single crystal silicon rods, as also micro-mechanical devices. 1990 market information for photovoltaic silicon solar cells show a market share of 44% for monocrytalline silicon, 25% for polycrystalline silicon, and 30% for amorphous silicon[7]. Present market trend also shows a decreasing demand for amorphous and non-silicon solar cells, whereas that for single crystal silicon is steadily increasing.

In semiconducting materials, these forbidden gap in between the conduction band and the valence band is small, which makes them suitable for photovoltaic effect. When solar energy falls on the cell, electrons flow from valence band to the conduction band resulting in flow of electric current.

Conduction band
Forbidden gap
Valence band

In insulators forbidden gap is too big to allow the above transition. Metals altogether have no forbidden gap and thus incident solar ray directly go to valence band of the atoms merely exciting the electrons thermally. Accordingly, metals do not show any photovoltaic effect and incident sunlight merely cause heating of the metal body.

Table 2. Purity requirement of silicon for electronic devices.
(Ref: Encydo. Mat. Sc. & Engg., Vol.6, Ed.M.B. Bever, Pergamon Press, NY (1986)

Impurity	ppm	Impurity	ppm
Group III elements	less than 0.3	Carbon	less than 300
Group V elements	less than 1.5	Oxygen	less than 50
Heavy metals	less than 0.1	All other impurities	less than 0.001

1.3 GENERAL USE PATTERN OF SILICON METALS

Both metallugical grade and high purity silicon have a number of commercial applications, summarized as follows:

(a) Metallurgical grade silicon is used as alloying element for strengthening metals like aluminium, magnesium, copper and a few others.
(b) Silicon metal has deoxidizing effect on steel and to some extent it imparts inertness to ferrous alloys.
(c) Steel with high silicon content is used in making core laminations for the transformer industry.
(d) High purity silicon is used in semiconductor devices, such as power rectifiers (for better heat dissipation), diodes, transistors, and photovoltaic solar cells. In certain applications in electronic industries, silicon is preferred over germanium, strictly for electrical reasons. However, greater interest in silicon in such industries is due to its greater temperature tolerance over silicon. Thus devices made from silicon can tolerate an operating temperature of about 180°C, while the upper limit for germanium is only 80°C.
(e) Sizable amount of silicon crystals is used in piezoelectric devices; thin wafers of quartz crystal control the frequency of radio oscillators by vibrating at a very exact frequency.
(f) Silicon of 99% purity is used as starting material for silicon resins, oils, and elastomers. Some of these materials are used as electrical insulation, mould release agent, for water repellant coating, hydraulic-fluid in polishes and lubricants, in cosmetic, and other applications.
(g) Organosilanes (such as CH_3SiHCl_2) which contain silicon-hydrogen bond, are converted into silicon polymers which retain the reactive silane groups. These special silicones are used to impart water repellant films in textiles, leather, and automotive windshield glasses.
(h) Silicon as its oxide is used in ceramics and glass industries, production of high temperature tolerant materials like silicon-carbide (abrasives), silicon nitride (for gas turbine blade manufacturing), etc.
(i) Because of electronic band gap, silicon normally does not emit light. But when doped with Er it emits light, and Er-doped silicon is used in microphotonics and integrated optoelectronics, as LED's and light emitters[8,9].

Still newer discoveries of special properties of its compound are taking place everyday, stretching the horizon of its use still further.

CHAPTER 2

Production and Purity

2.1 PRODUCTION OF METALLURGICAL GRADE SILICON

Metallurgical grade silicon (with silicon content of 98-99%, the rest being iron about 1%, aluminium 0.5%, phosphorus 0.17%, boron 0.01-0.001%, titanium 0.11%, manganese 0.16%, calcium 0.4%, approximately) is being prepared from silica by carbothermic reduction:

$$SiO_2 + 2\,C = Si + 2\,CO \quad (\Delta H = 695\text{ kJ})$$

The process is carried out in an arc furnace and electrical energy required is 12.5-14 MWH/ton of silicon. Yield of metallurgical grade silicon by above reduction process is about 1 ton silicon from 3 ton of quartzite. The other reaction:

$$2\,SiC + SiO_2 = 3\,Si + 2\,CO$$

also takes place at 3100 F, as SiC is also formed to some extent in the mixture. In an effort to increase the yield of silicon by preventing its vaporization and by minimizing formation of SiC, several silicates have been added in varying ratios to the silica-carbon charge. To remove carbon from low grade silicon, it is remelted in a crucible which has an alkaline earth carbonate at the bottom. On decomposition of the carbonate, CO_2 bubbles through silicon and reacts with carbon to form CO gas. Using this method silicon with a resistivity of about 0.5 ohm can be prepared. Commercial furnaces producing metallurgical grade silicon by this process, consume about 10-30 MW of electrical energy to power carbon electrodes 1 m in diameter. These furnaces can produce about 20,000 ton of silicon per annum. Besides the above reaction, other reactions also occur within the furnace, but to a much lesser extent; these are—

$$SiO_2 + C = SiO + CO$$
$$SiO + 2C = SiC + 2CO$$
$$SiC + SiO_2 = Si + SiO + CO$$

The primary by-product is gaseous SiO, which is collected, oxidized to SiO_2 and used as fumed silica for catalytic applications, as an additive in paints, as an extender in rubber and silicone products.

Quality of these crude silicon produced, plays a crucial role in subsequent refining steps. With demand of higher purity and surface quality in high-purity silicon growing with time, demand for better quality metallurgical grade silicon is ever increasing. Purity of these metallurgical grade silicon also affects performance of catalyst during its subsequent refinement steps as well. For example, in the production of intermediate product trichlorosilane (TCS) for silicon purification, presence of impurities drastically reduces production yield. Copper which is a versatile catalyst in silicon chemistry[10], its catalytic effect shows a main pathway towards TCS (trichlorosilane) and silicon-tetrachloride formation as well as a side reaction towards dichlorosilane, branching off the main route during early stage of reaction between metallurgical silicon and HCl gas[11]. These branching off is facilitated by impurities in the metallurgical silicon. Thermodynamic and kinetic studies with Ni-Si-Cl-H system[12] shows that transition metal silicides formed from impurities have pronounced effect on hydrodehalogenation of $SiCl_4$ and TCS. Formation of silicide phase in $NiSi_x + NiCl_2$ has been studied by above investigators and noted formation of $SiCl_2$ phase by fast migration of silicon in the bulk silicide in presence of transition metals, which in-turn brings about a phase transformation in the bulk silicide. Similarly, Wakamatsu *et al*.[13] at Tokuyama Corporation, Japan, noted that the main reason in drop of production of TCS in silicon-HCl reaction was due to coexistence of impurities like iron and phosphorus, which ultimately destroys the selectivity of copper catalyst towards TCS formation and hence drop in production yield with time. However they found that presence of aluminium can restore to some extent this loss in selectivity caused by iron and phosphorus.

This understanding regarding the role of impurities in crude silicon, has led to refinement in the process of its production and has recently been able to bring the impurities to tolerable limit[14]. This in-turn has boosted production of metallurgical silicon to a great extent. For example, Larochelle *et al*[15]. at SKW Canada Inc (Becancour, Canada) through systematic work on influence of the arc in separating impurity in metallurgical silicon, developed an injection technology that resulted in a remarkable increase in silicon metal production in their furnaces in the year 1997 and the output further grew by 7% in 1998. New developments[16] in prebaked electrode technology also has improved performance in this direction and allowed silicon manufacturers to operate at high productivity rate with low impurities. Electrode losses from thermal stresses and oxidation was also curtailed, resulting in increase in production effeciency and economic gain. New generation electrodes claimed to lower oxidation by 50% while substantially reducing thermal stresses. Oxidation rate improvement was affected by incorporating phosphorus containing inhibitors to the electr ode, which apparently do not contaminate the melt. Thermal stress improvement on the other hand, was gained from new electrode formulations.

Raw material silicon-dioxide used in these processes also plays a vital role in ensuring final product purity. Prokhorov *et al.*[17] at General Physics Institute Moscow, Russia, studied a large number of quartz samples all over the world, that is being used in carbothermic reduction for production of silicon and suggested best qualities available at different mine sites. Recent market survey[59] shows that bulk and surface quality of most of the high purity polysilicon available in the market from various manufacturers are in good agreement with respect to impurity level, while in contrast, surface and bulk impurity level in granular polysilicons are still not competitive compared to chunk polysilicon.

Based on thermodynamic equilibrium between silicon and refining slag (as is done in steel industry for steel refining), molten metallurgical silicon has been refined in the furnace itself.[19] Experimental results indicate, slag treatment in silicon refining, brings down calcium activity down to 0.00059 ± 0.00002% at 1823 K. Influence of slag viscosity on such refinement procedure has also been considered by above Brazilian investigators. Studies with alternative fluxes like Na_2O, K_2O, and MgO at 1773 K indicate that, magnesium content lower than 200 ppm could be obtained when 10% of MgO is added to a (80% SiO_2 + 22% CaO + 18% Al_2O_3) slag. Sodium contamination with 10% Na_2O slag (50% SiO_2 + 30% CaO + 20% Al_2O_3) was not higher than 60 ppm. Different cooling rate brings about significant change in structural index (a micro-structural parameter) in such a case. Structural index has bearing on grain size of silicon during atomization or preparation of granulated silicon.

Another plant in Brazil, Electrosilex, reported refining metallurgical silicon using reductants like charcoal and wood-chips. The refining process was carried out in the pouring ladle, in order to control calcium and aluminium concentration by simultaneously blowing with a gaseous mixture of nitrogen and oxygen through a porous plug. The blowing process was controlled by a PLC (Program Logic Controller) and both the composition and viscosity of the slag maintained at a predetermined level, in order to remove impurities to the maximum level possible. The predetermined values were calculated by a software computer program using thermodynamic values available in the literature. Results confirm that proper use of synthetic slag is the best way of controlling impurities in the molten silicon bath.·Kawasaki Steel Corporation, Japan, recently developed[21] an improved graphite container in silicon manufacture, which can withstand severity of the thermal gradient and prevent cracking and consequent contamination possibility which often occur in such industries. The container developed by the above company has a specific shape and covered with a graphite layer (serving as an insulator) and a back-up container also made of graphite plates, in order to prevent molten metal contamination in case of any crack.

2.2 PRODUCTION OF HIGH PURITY SILICON

One of the mothods of producing high purity silicon from silica, is to convert the latter to SiF_4, and then react it with metallic sodium:

$$SiF_4(g) + 4\,Na(1) = Si(powder) + 4NaF(solid)$$

One of the possibilities of deriving SiF_4 from silica would be reacting silica with UF_4, which do not contaminate the final product SiF_4 with radioactivity. Alternatively SiF_4 can be obtained from phosphoric or super-phosphate making plant, where silica is added to rock-phosphate and sulphuric acid mixture to facilitate phosphate separation. This addition of SiO_2 forms fluorosilicic acid, which can be precipitated by alkali to form sodium-fluorosilicate. Further addition of strong sulphuric-acid to this precipitate generates[22,23] SiF_4. The exothermic nature of above SiF_4 and sodium reaction, provides sufficient heat to maintain the process at 500°C. To separate silicon from sodium-fluoride (NaF), the product mixture is first melted to form two immiscible phases for which liquid-liquid extraction (migration of impurities from silicon to sodium-fluoride) provides additional purification. This low cost silicon manufacture by Stanford Research Institute process[24], is possible only if the by-product NaF can be sold profitably, since the reaction yields four times as much NaF as silicon. Silicon obtained by this process has purity in the range of 98-99 %.

High purity silicon necessary for making solar cells or for semiconductor devices, calls for purity of the order of 99.999999 % or impurity level in few ppm or preferably in ppb level. Such an extraordinary class of purity can be achieved from metallurgical or crude silicon, by changing the phase of silicon or producing an intermediate chemical. This is achieved by converting metallurgical silicon to trichloro-silane (in short TCS) or silicon-tetrachloride ($SiCl_4$) which later is thermally cracked to obtain the desired purity level. TCS method of producing high purity silicon is the main pathway for producing commercially high purity silicon and we will discuss this method in detail in following section.

(a) Trichlorosilane method of producing high purity silicon

Almost all high purity silicon produced world wide comes from trichlorosilane mode of production, except a few which rely on silane (SiH_4) or $SiCl_4$ route. The starting material for this route is metallurgical grade silicon, described earlier and produced by carbothermic reduction of silica. In this process the metallurgical silicon is reacted with dry hydrochloric acid gas in a fixed or fluidbed reactor (Fig.2).

$$5\ Si + 16\ HCl = 4\ SiHCl_3 + SiCl_4 + 6\ H_2 \quad \Delta H = -295\ \text{kcal}$$

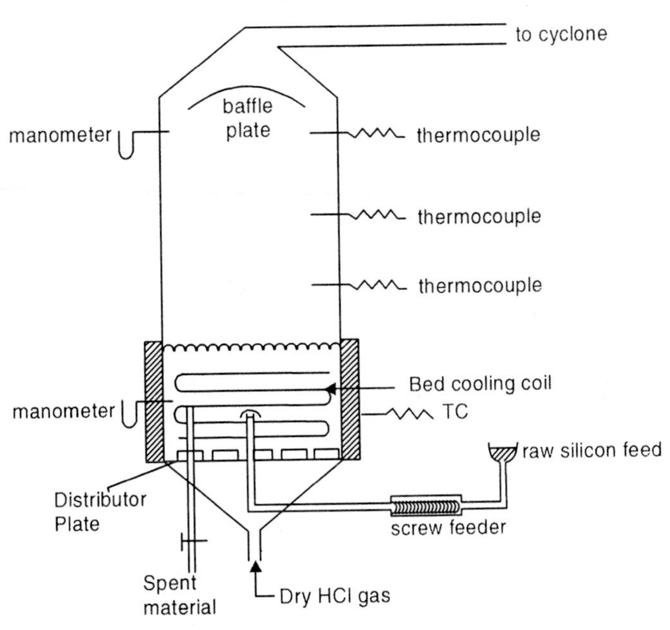

Fig. 2: Schematic diagram of a fluidbed reactor for TCS production.

Material balance and block diagram of a possible automatic controlled silicon plant using TCS is shown in Table 3 and Fig. 3 below[25].

Table 3. Mass balance of a TCS process[25].

Crude silicon (metallurgical grade)	143 g
Dried hydrochloric acid gas	584 g
Trichlorosilane produced	135.6 g
Silicon-tetrachloride	170.1 g

As may be noted in Table 2, at equilibrium a good amount of silicon-tetrachloride is also produced as a by-product along with the main product TCS. Since the reaction involves hydrochloric acid gas, the plant will be most conveniently located near a caustic-soda producing plant, whose by-product is hydrogen and chlorine gas. These two gases from the chloro-alkali plant can be burnt through a suitable burner to produce hydrochloric acid gas under pressure. Alternatively, in remote location, transported concentrated liquid hydrochloric acid may be added dropwise to oleum or

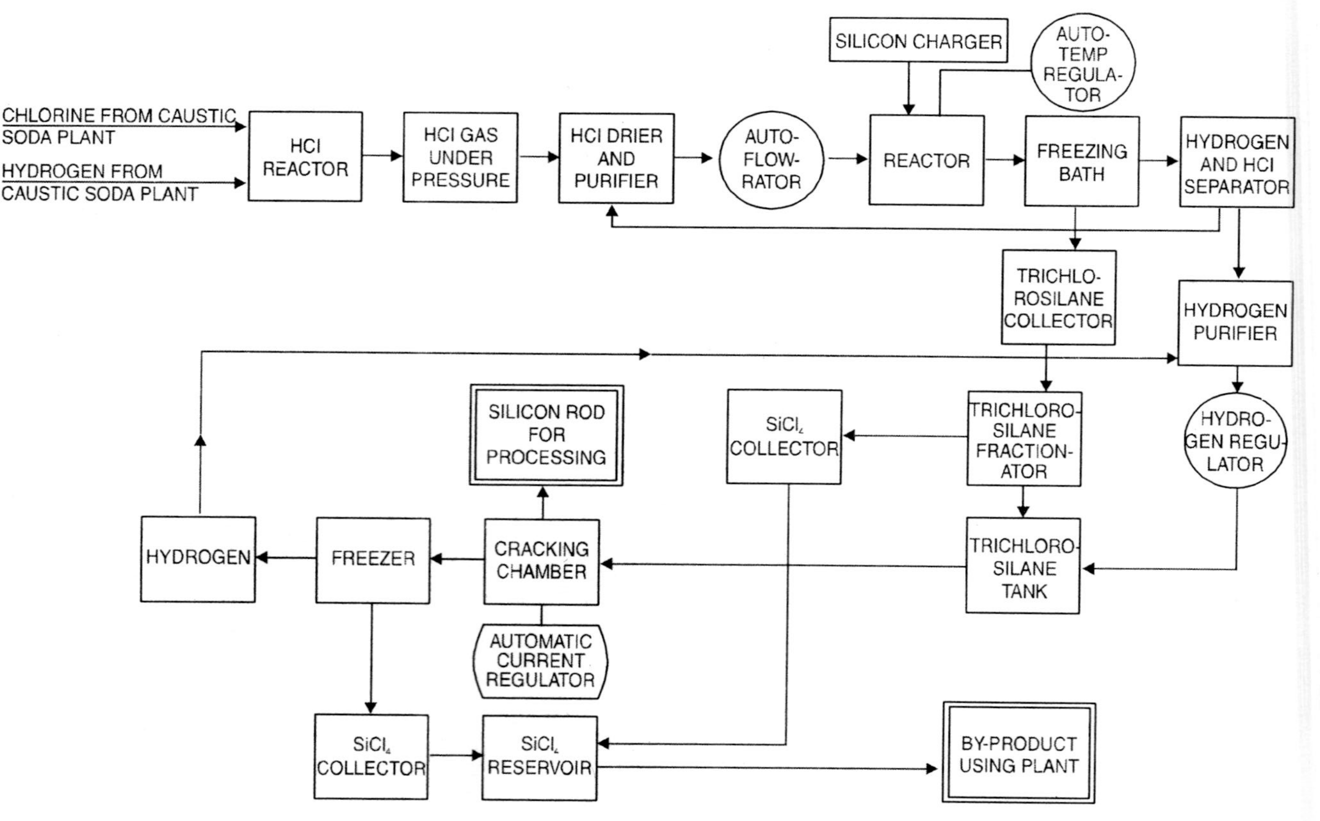

Fig.3: Block diagram of a complete automatic silicon production plant.[25]

fuming sulphuric acid in a suitable gas-generator to produce hydrochloric acid gas. In later case, 5 litre of concentrated liquid HCl and 6 litre of concentrated sulphuric acid wll produce 1200 litre of hydrochloric acid gas at 25°C. The final sulphuric acid concentration, or spent acid concentration will be 68 % H_2SO_4 containing 0.37 % HCl.

Accordingly, later method of HCl gas production will call for utilization of large volume of spent sulphuric acid everyday, to make the process economical. One recourse to this direction would be to recharge the spent acid with SO_3 gas[26] and reuse the regenerated acid. The spent acid can also be concentrated by special distilling apparatus[27]. Part of the acid may also be sold to hardware store for domestic cleaning purpose. Dilute sulphuric acid can also be economically electrolyzed to produce hydrogen and oxygen gas (without AC/DC converter) using GaAs photocell exposed to sunlight with an overall efficiency of 12 %. Yet another way of recovering sulphuric acid from spent solution would be to treat the acid solution with iron-oxide to form solid precipitate. The precipitate is then heated to 550-750°C to give out SO_3 gas. The residue again heated to a higher temperature of 800°C (this step regenerates iron-oxide for reuse). SO_3 thus generated can be used to recharge the lean acid solution for reuse[28].

The reaction mentioned above between silicon and HCl, besides producing TCS and $SiCl_4$, also produces a small amount of dichloride (SiH_2Cl_2). This dichloride has a boiling point of 8°C and can be converted to TCS by a process involving reaction with HCl at 0-75°C in presence of activated carbon (hydrogen used as carrier gas)[29]. This dichloride in any case can also be thermally cracked to get.pure silicon.

Properties of trichlorosilane are presented in Table 4.

At Wecker-Cheme GmbH, Germany, TCS is produced in fluidized bed of silicon particle[30] with a diameter less than or equal to 50 µm, heating not less than 15 hour at 100-350°C in air or nitrogen before inserting MeCl with HCl into the reactor. The process gives out in one batch 60 g silicon residue (diameter equal to or less than 50 µm) after the reaction at 120-140°C for 70 hour. These flying residue are caught in a cyclone separator and mixed with 60 g fresh material into another vertical glass tube at 320°C and fluidized with 40 g HCl for 4 hour to get additional TCS. TCS mixture thus generated was 400 g in total from above masses and had a composition of TCS = 95 %, $SiCl_4$ = 4.9 %, and SiH_2Cl_2 = 0.1 %. TCS and $SiCl_4$ can also be prepared continuously in fluidized-bed by reaction of silicon with HCl at 260-1200°C. A cone within the fluidized-bed with its base resting on the approach flow bottom, also serves as a cooling surface and nonreacting substances as well as solid reaction products are removed continuously from below the cone. Temperature plays a decisive role in TCS : $SiCl_4$ ratio in throughput gas by this process. In other attempts to produce TCS in fluidized-bed[31] carrier gas like inert nitrogen, argon, or

Table 4. Properties of Trichlorosilane (TCS) 7, 141

Property	Value
Character	Colourless fuming liquid
Melting point	– 134°C
Boiling point	36.5°C
Refractive index[25] (n_d)	1.3983
Viscosity at 25°C (centi-stoke)	0.23
Vapour pressure (mm): at –54°C	10
at –26°C	60
at 14.6°C	400
Heat of vaporization (cal/g)	46.7
Flash point, Cleveland Open Cup (F)	– 18
Auto-ignition temperature (F)	220
(Vapour highly inflammable and explodes in contact with hot surface)	
Specific heat	0.23
Coefficient of expansion (per°C)	0.0019
Solubility	Soluble in organic solvent Decomposes in water & alcohol
Heat of formation (kcal/mol)	– 40
Storage	It does not react with glass or steel & accordingly stored in such container using hydrgen as sealing gas.
Price (1980 cost)	$0.80/1b technical grade & $2.00/1b pure grade.

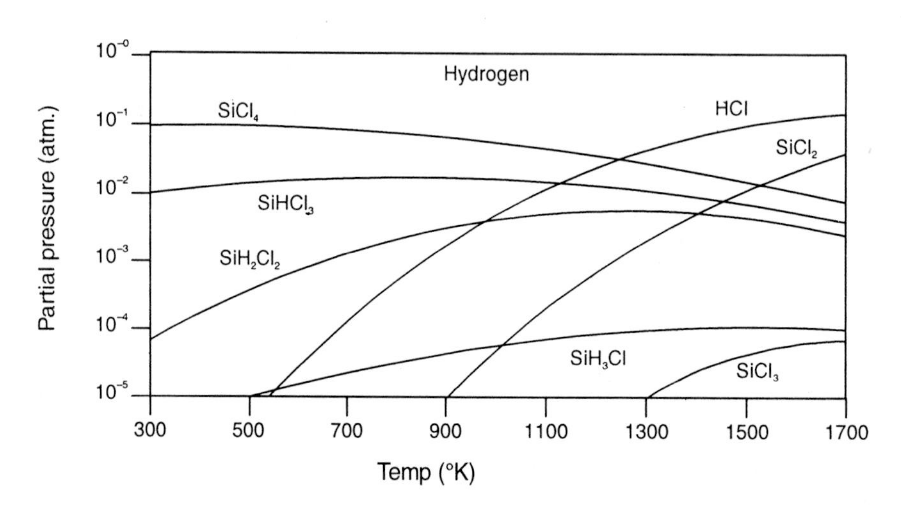

Fig. 4: Equilibrium concentration in Si-Cl-H system for Cl/H = 0.1 and P = 1 atm. (Ref: Sirtl E., Hunt L.P., Sawyer D.H, *J Electrochem. Soc.* 121 (1974) 919.

helium, or reactive gases like chlorine and $SiCl_4$ has been used as heat convector as well as fluidizing gas. Hydrochloric acid gas fed in pulses is also reported to enhance yield of TCS. Higher temperature in the bed followed by rapid cooling, also is reported to increase TCS yield 1.8 fold. Quenching in the later case is affected by spraying at top of the bed with by-product liquid $SiCl_4$ at 20°C. Depending on the cooling rate, the $SiCl_4$: TCS ratio can be adjusted between 4 : 1 and 0.2 : 5. Besides copper, the conventional catalyst, aluminium, nickel and antimony-pentachloride has also been used as a cocatalyst with consequent lower reaction temperature and boost in TCS yield by 21 %. In fixed bed reactor where silicon particle size of 3 mm and down is used[25], the reaction is carried out isothermally at 260-280°C. Fluidized bed is generally at higher temperature than this, depending upon product preference of TCS or $SiCl_4$ in the range 260-1200°C.

During production of TCS in a stirring or agitated bed, it has been found that screw type stirring vane gives best result. In these reactors flow of hydrochloric acid gas is 6-10 times as great as static bed. Metallurgical silicon of size less than 2 mm is used in such reactor, with 5-10 % copper powder (100 mesh) added as catalyst. Maximum yield with such reactor is found in the range 280-290°C; reaction does not start till 270°C, and above 300°C TCS production drops abruptly. Increasing copper catalyst concentration above 10 % does not change reaction rate. Stirring rate of 300 revolution/min generates maximum TCS (with about 90 % conversion). Condenser of freon operating cycle cooling in –40°C range is used to trap the produced TCS vapour.

Extremely fine particles of silicon (50-800 μm) obtained by atomizing molten silicon in nitrogen atmosphere, are claimed to give a high yield of TCS. A screw conveyor has also been used for producing TCS from ferrosilicon. In yet another innovative process, silicon is first reacted with $SiCl_4$ at 1100-1300°C and the resulting product is treated with hydrochloric acid gas. The conversion of $SiCl_4$ to TCS is reported to be 50-60%. Adding $SiCl_4$ to HCl, prior to contact with powder silicon is claimed to enable complete conversion to TCS. A similar process has been described by Oda[32]. recently, for production of TCS involving reaction of metallurgical silicon with hydrogen and $SiCl_4$, chilling the reaction product to separate out TCS and $SiCl_4$, reacting the pure TCS with hydrogen, to produce high purity polycrystalline silicon, and recycling back the $SiCl_4$ containing by-product gas. The same investigators used carbon based absorbent layer for supplying hydrogen, and the product seems to contain very little carbon, within tolerable limit[33].

Metallurgical grade silicon can also be reduced with magnesium-silicide (Mg_2Si) to form silane (SiH_4) which eventually can be thermally cracked to produce pure polycrystalline silicon. Magnesium-silicide, in-turn, can be produced by reduction of silica with magnesium metal beyond its stoichiometric ratio (under NaCl cover):

$$SiO_2 + 4\ Mg = Mg_2Si + NaCl$$

Acharya *et al.*[34] (IIT, Kharagpur, India) reported metallothermic reduction of acid leached rice husk with magnesium followed by acid leaching at a much lower temperature compared to that required by other conventional methods, to produce polycrystalline silicon (or 10 μm size) with sufficient purity (above 99 %). This crude silicon can be further purified (e.g. by reactive gas blowing) to produce solar grade silicon at a cost of $ 1.2/kg (1979 price level), and the total energy requirement per kg of material worked out to be 110 kWH/kg solar grade silicon. Cost estimate and process flow diagram of this technology is shown in Table 5 and Fig. 5.

Table 5. 1979 Cost Estimate of Si production[34].

Basis:			
	Plant capacity	= 1000 ton solar grade Si/year	
	Operation time	= 7200 hour	
	Average labour cost	= Rs. 2.20/hour	
	Power consumption	= 110 KWH/kg of Si	
	Power cost	= Rs. 0.20/KWH	
	Investment	= Rs. 400 lakh	
Material:			
	Rice husk	10,000 ton	Rs. 2 lakh
	Sulfuric acid	1,000 ton	Rs. 10 lakh
	Hydrofluoric acid	500 ton	Rs. 10 lakh
	Hydrochloric acid	2,000 ton	Rs. 20 lakh
			Total = Rs. 42 lakh
Utilities:			
	Electricity	110×10^6KWH	Rs. 220 lakh
	Process water		Rs. 15 lakh
			Total = Rs. 235 lakh
Labour:			
	500 persons per 7200 hour (@ 2.25/hour)		Rs. 82 lakh

Interest: 19% interest on fixed capital of Rs.400 lakh = Rs. 76 lakh
Total Cost = Material + Utility + Labour + Interest = Rs. 434 lakh
Production Cost = Rs. 43.4/kg of solar grade Si

Prasad[35] also suggested a very low solar grade silicon price ($ 1.2-1.8/kg at 1979 price level) under Indian conditions, by fully utilizing the by-product $SiCl_4$. Thus a good part of sale value comes from by-product sale.

Silane (SiH_4) has been produced industrially by reacting above magnesium-silicide with ammonium-chloride and ammonium-bromide in liquid

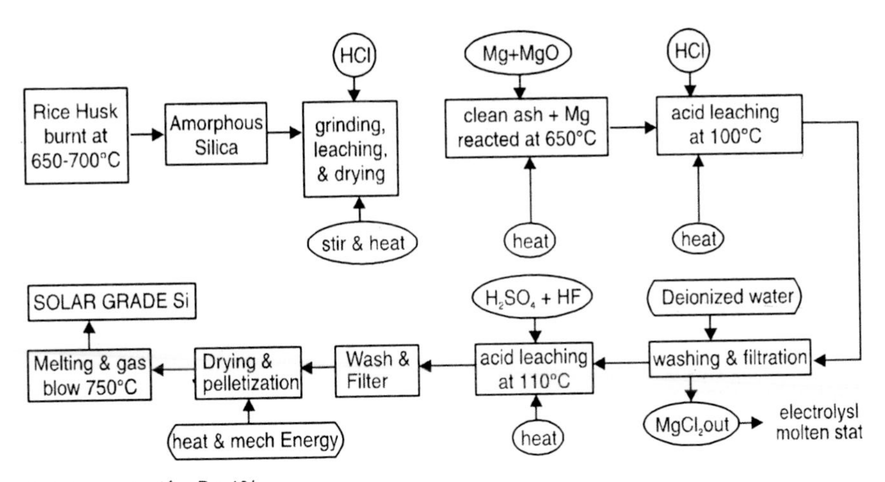

Fig. 5: Process diagram

ammonia solvent. From this process, it seems that the Mg_2Si route of silane production will be economically viable, if the by-product like $MgCl_2$ etc. can be recycled in the process.

Very high purity Trichlorosilane (TCS)

Very high purity TCS can be produced from commercial grade TCS by reacting it with thioglycolic acid, β-naphthylamine, and salts of ethylenediamine-tetra-acetic acid. An extremely pure product is also obtained by extraction with methyl-cyanide. Other methods include, adsorption of the impurities on columns of activated silica, activated carbon, ion-exchangers, titanium sponge, or by treatement with acetals or salt-hydrates, which cause partial hydrolysis of the impurities. Generally boron impurities are first converted from volatile-halides or hydrides to non-volatile acids or oxides (e.g. by passing moist nitrogen). Phosphorus generally exists in these solutions as PCl_3 and can be removed by MnO_2, which forms oxychloride having a much higher boiling point than PCl_3.

By-product tetrachloro-silane ($SiCl_4$)

As may be noticed in the reactions mentioned above with respect to TCS production, substantial amount of silicon-tertrachloride is produced as by-product along with the targeted material TCS. Accordingly, to improve economics of TCS production, the by-product value needs to be realized through some uses. In the following, we will discuss some such possibilities. Properties of silicon-tetrachloride are presented in Table 6 below; since in dry condition it does not attack steel, it can be stored and transported safely in SS-container.

Table 6. Properties of Silicon-tetrachloride

Property		Value
Character (thermally stable but chemically very reactive)		Colourless fuming liquid
Melting point		–69.4°C
Boiling point		57.3°C
Refractive index (n_d^{25})		1.4146
Specific gravity (at 25°C)		1.4736
Viscosity (at 25°C)		0.35 centi stoke
Vapour pressure:	at –34.4°C	10
	at 5°C	100
	at 38.4°C	400
Heat of vaporization (cal/g)		40.3
Heat of formation (kcal/mol):	for liquid	153
	for gas	145.7
Coefficient of expansion (per°C)		0.0011
Specific heat		0.20
Solubility:		Soluble in organic solvents like benzene, chloroform, and petroleum-ether. Decomposes in water and alcohol; it forms esters of silicic acid with alcohols, and partial hydrolysis with water-ether mixture produces silicon-oxychloride with general formula $Si_nO_{n-1}Cl_{2n+2}$.

Commercially exclusive production of silicon-tetrachloride is carried out by the exothermic reaction between silicon-carbide and chlorine. The SiC used is generally a by-product material and not suitable for abrasive purposes. A new high temperature process involving clay has been developed, but no results have been published yet.

Uses of Silicon-Tetrachloride

(A) Production of Silane (SiH_4)

Silane can be produced from silicon-tetrachloride by reacting it with a metal-hydride (Standard Telephone & Cable, USA, process). The most common hydride used is $LiAlH_4$, and the reaction involved is as follows:

$$SiCl_4 + LiAlH_4 = SiH_4 + LiCl + AlCl_3$$

The hydride is dissolved in ether and the silicon compound is bubbled through the solution. Hydrides of boron, arsenic and phosphorus are removed by passing the silane through activated charcoal trap at about 80°C and between 0 and 30°C. Removal of boron is more effective if water vapour is added to the gas, which was then stored for not less than 24 hour at room temperature, before cooling at –80°C to get rid of arsine and phosphine. The purified silane can be decomposed on a silicon seed heated at 1050-1150°C, after being diluted with inert carrier gas, so that the silane pressure reduced to between 8 and 17 mm Hg. Other metal hydrides may also be used for reducing $SiCl_4$, while alternative methods for purifying the product, including gas chromatography has been suggested[36]. To lessen the hazards of use of silane, a less volatile solvent (tetraethylene-glycol) is used for the reaction and thus limit the vapour pressure of the silane to 9 mm by means of an isopentane bath kept at –160°C, which liquifies excess silane. The volume of liquid silane is also limited in the process, by controlling the rate of addition of $SiCl_4$ to the solution or slurrying excess $LiAlH_4$ in the solvent.

Again like TCS, high purity silicon can be produced by pyrolytic decomposition of silane[37]. This technique of dissociating silane by pyrolysis has nevertheless been adopted by a handful of manufacturers around the world (especially in Japan) due to detonation problem with this gas. Silane explodes and ignites in air, and is decomposed by water containing traces of alkali. It reacts explosively with halogens to give silicon halides, and in the absence of air it decomposes to silicon and hydrogen at 400°C. Nevertheless, silane is a stable hydride and also has some special attractions for producing high purity silicon through its decomposition. For example, it is a gas and therefore can be easily purified and kept purer than a liquid or solid. Material of construction is of lesser importance in case of silane, while the decomposition temperature is too low for container contamination and saves energy[38] (nearly half of that required for TCS).

Union Carbide, Moses Lake, Washington, USA, produces 2400 ton/year silicon by its proprietary process involving reaction between metallurgical grade silicon with $SiCl_4$ and hydrogen to produce TCS. TCS then catalytically disproportionates to dichlorosilane to get silane (SiH_4). One disadvantage of this process is high recycle rate; one pass converts only 30 % for the first two steps and 11 % for the last one. Ethyl Corporation (Richmond, VA, USA) claimed a better energy saving silane process, which has high reaction yield and small stream flows without re-cycle.

The new single step (simpler, cheaper and cleaner) process for making silane involves, reduction of SiF_4 with sodium-hydride in boiling solvent such as diphenyle-ether (which boils at 259°C). Chlorosilane can also be used, if need be, as raw material for manufacture of silane (SiH_4). A previous method of producing silane has been a multi-step reduction of $SiCl_4$ with $LiAlH_4$, which proceeds through intermediate trichloro–, dichloro–, and chloro– silane, or sequential disproportionation of TCS through dichlorosilane. The new process with SiF_4 was tested at 40 ton/year SiF_4 input or 10 ton/year SiH_4 output capacity in commercial scale successfully. Recently Hattori *et al.*[39] of Tokuyama K.K, Japan, has described a new quaternary ammonium salt based catalyst for producing silane by disproportionating halosilanes.

Silane has also been produced in high yield by heating silicon at or above 800°C (below its melting point) for not less than 10 min with magnesium to form an alloy with mechanical kneading process and then reacting the alloy with ammonium halides in liquid ammonia solvent[40]. By controlling the alloy firing temperature and time, formation of disilane and its yield can also be controlled by this process.

Besides silanes, monosilanes has also been produced by—reducing SiF_4 with calcium-hydride, by treating siloxene ($Si_6O_3H_6$) with ammonia at temperatures below –33°C, by reducing silica or silicates with hydrogen in the presence of aluminium and aluminium-halides at 175°C and 400 atmospheric pressure, by disproportionation of triethoxy-silane in the presence of sodium, and disproportionation of chlorosilane in contact with a catalyst containing anion exchange resin loaded with group-VIII metals[41] (claimed to be an efficient process).

(B) PRODUCTION OF PURE SILICON BY REACTION OF $SiCl_4$ WITH ZINC:
The interaction of zinc and $SiCl_4$, both in vapour phase, was the first method to yield reasonably pure silicon. Zinc is a very suitable reductant. It is available with good purity and has a negligible solubility in silicon. It does not form a silicide, while both it and zinc-chloride formed in the reaction are readily vaporized. Zinc is reasonably cheap. The process as developed by Dupont Co, USA consists of distilling zinc into a reactor of fused quartz kept at 950-1050°C, where it meets with a slight excess of $SiCl_4$ which has been vaporised and heated to 650°C. The reactor contained baffles to promote the reaction. Needles of silicon were deposited while excess reactants and zinc-chloride passed into a collector; unused $SiCl_4$ was separated and recovered. In a later development, the reaction chamber was divided into two zones, a smaller pre-reaction zone where 5-25 % of the silicon is deposited together with most of the impurities. Silicon in a purer form is then produced in the second, larger zone. The product is treated with hydrochloric acid and after washing and drying, is heated with a mixture of hydrofluoric-acid and sulphuric acid. After being washed free of acid with distilled water, it is finally filtered and dried. The estimated purity of silicon by the original method is reported to be 99.97 % and has a resistivity of 0.01 ohm cm.

Recently Natsume and Kumanao[42] at Osaka Titanium Co, Japan, patented a closed cycle high purity silicon manufacturing technology using $SiCl_4$ and zinc. In this process, liquid or gaseous $SiCl_4$ is reduced with molten zinc to produce polycrystalline silicon and zinc-chloride. Zinc-chloride separated from polycrystalline silicon, is electrolyzed to form back elemental zinc and reused again as reducing agent in the process, while chlorine is reacted with hydrogen to produce hydrochloric acid gas, which has been used to react with metallurgical grade crude silicon to form TCS. Thus production of pure polycrystalline silicon, recovery of by-products and its reuse, is carried out in a closed cycle for better economy of the process as well as to avoid environmental pollution. Silicon produced by this process is claimed to be of solar grade.

(c) Production of pure silicon by reaction between $SiCl_4$ and sodium:
This is the earliest method attempted to produce high purity silicon. The reaction was carried out entirely in vapour phase at 440-460°C under reduced pressure. The product is a microcrystalline silicon powder. Resistivity of the pressed powder sample and of blocks produced by melting was reported to be between 10^2 and 10^4 ohm cm at room temperature.

(d) Production of ethyl-silicate from $SiCl_4$:
Ethyl silicate can be produced by reaction between $SiCl_4$ and ethanol (generated by fermentation process from molasses), according to following reaction:

$$C_2H_5OH + SiCl_4 = (C_2H_5)_4 . SiO$$

Ethyl-silicate as such do not have any binding power, but when treated with water it hydrolyzes forming a gel. Ethyl-silicate thus used with the required amount of silica to serve as binder, which after firing forms ceramics and refractories with excellent physical and chemical properties. Thus ethyl-silicate has provided binding of highest quality refractory aggregates, such as silliminite, mullite, alumina, zircon, zirconia, silicon-carbide, magnesia, in order to form shapes of the highest refractoriness. Degradation of the gel at elevated temperature results in progressive removal of organic residues and forming 'siloxane' groups together with active sites on the silicon-oxygen system. A ceramic bond is formed during this process[43]. Tetraethyl-silicate is also hydrolyzed slowly in water with little acid and the partially hydrolyzed tetraethyl-silicate forms gel or viscous product, which are used in silicate-paints for sealing and moisture proofing stone. With water these gels hydrolyses to form SiO_2, which is water repellant and blocks the pores of stones. In a similar way, tetraethyl-silicate with sand yields a heat resistant paste which can be used for the manufacture of molds and cores in metal casting industry. Silicic acid esters cement or putty, is made by mixing finely ground quartz, sand, or refractory brick with tetraethyl-silicate, an alkaline material and alcohol. The mixture sets to a solid mass mainly for production of small refractory articles and for cementing together electric heating elements. A mist of finely divided and very pure silica, can be prepared by combustion of tetraethyl silicate vapour with oxygen. This mist can be deposited on inner surface of incandescent bulb to improve diffusion of light (0.1-1 μm thickness); such light diffusing coatings are very cheap to produce.

(e) $SiCl_4$ vapour deposited steel:
National Physical Laboratory (New Delhi, India), Magnetic Material Division, has developed a process[44] for vapour depositing $SiCl_4$ on low silicon

steel sheet (the Si concentration being 1.5 wt %), and increased it up to 6.5 wt %. The product showed following property—B_s (1.5–1.8 T), H_c = 2–4 T/m, and electrical resistivity 90–100 micro-ohm cm.

(F) PRODUCTION OF HIGHLY DISPERSED (PRECIPITATED) SILICA FROM $SiCl_4$: Highly dispersed silica is produced by hydrolysis of $SiCl_4$. Water needed for the hydrolysis is produced *in-situ* by burning hydrogen (mixed with $SiCl_4$) in order to obtain a highly dispersed powder:

$$SiCl_4 + 2\,H_2O = SiO_2 + 4\,HCl;\ 2\,H_2 + O_2 = 2\,H_2O$$

The gas mixture ($SiCl_4$ + H_2) is flame hydrolyzed by a burner nozzle and the product silica in the form of aerosol is separated from its mixture with hydrochloric acid by a cyclone or filter. Leftover hydrochloric acid (small quantity) is further removed by treatment with steam and air in a fluidized bed reactor. The released hydrochloric acid gas formed is scrubbed with water and dilute hydrochloric acid is obtained as a by-product in the process. Commercial producers like Wacker-HDK, Degussa-Aerosil, and Tokuyama Soda's Rheosil variety, control the properties of their product pyrogenic silica further, by varying reaction parameters, such as flame composition and flame temperature. These commercial products have surface area in the range 50-400 m^2/g with bulk density less than 20 g/litre. While packing, these powders are pneumatically transported and tapped density increased to 50-120 gm/litre by compacting rollers or vacuum packers.

Raw material for the production of 'precipitated silica', on the other hand, are aqueous alkali metal silicate solution (e.g. water glass) and acids, generally sulphuric acid. Silica gel is produced in acidic condition, but in case of precipitated silica precipitation is carried out in neutral or alkaline media. Stirring and heating prevents formation of a gel structure. Filtration is done in a rotary filter, belt filter or filter press. These products have a BET surface area in the range 80-700 m^2/g, and silica content 98-99.5 %, Na_2O content 1-0.2 % and Fe_2O_3 content 0.03 %. Its current worldwide use pattern is—shoe soles (40 %), tyres (16 %), carrier-free flow (16 %), toothpaste (9 %). In addition colloidal silica finds use in treatment of textile fibers, ceramic binders, protective spraying agent for foundry moulds (which increases service life of moulds and gives improved surface finish to cast product), surface treatment of paper, board and cellophane by increasing coefficient of friction[45].

A process for silanized colloidal silica for application in density gradient media, as an aid in separating celloids, as an anti-abrasion coating, and making toner material have been described recently by Van *et al*[46]. These materials are prepared by aqueous suspension of organosilanized colloidal silica particle prepared from aqueous silanes (organo) and colloidal silica.

(G) FUMED SILICA FROM $SiCl_4$:
Fumed silica is a synthetic amorphous silica produced by burning $SiCl_4$ in an O_2–H_2 flame[47]. Surface area of these fumed silica is in the range 50-400 m^2/g. Using particle sizing technique, fumed silica shows micron sized particles leading to surface areas markedly lower than expected. Fumed silica appears as a fluffy solid with bulk densities greater than or equal to 0.03 g/cc, being invariant over the wide range of the surface area. However, performance of fumed-silica in technical applications, such as its thickening efficiency in fluids, mainly fail and the reason remain unambiguous. $SiCl_4$ is also used as 'smoke screen' in combat or for military applications.

(H) PRODUCTION OF SILICON-NITRIDE FROM $SiCl_4$:
Silicon-nitride possesses some extraordinary property like very high temperature withstanding capacity, low thermal expansion and refractoriness of high order, inertness towards most of the chemicals even at high temperature, etc. Accordingly silicon-nitride nowadays finds extensive application in high temperature engineering ceramics and in the form of whiskers as reinforcing material in cermets. Jeffers and Bauer[48] produced silicon-nitride by reacting $SiCl_4$ with a mixture of hydrogen and nitrogen, which yields amorphous powder of silicon-nitride:

$$3\, SiCl_4 + 2\, N_2 + 12\, H = Si_3N_4 + 12\, HCl$$

Silicon whiskers are being prepared by Gosai Chemical Co, Japan[49] by chemical vapour deposition route, using hexachloro-disilane with ammonia at 950-1700°C.

Direct nitridation of granulated silicon powder in fluidized-bed in the non-oxidizing atmosphere of nitrogen or ammonia at 1000-1400°C also produces silicon nitride. The amount of heat removed by this process is controlled based on the relationship between the particle size of silicon powder and the superficial linear velocity of the fluidizing gas. In such process, slugging is avoided and the process gives out uniform powder of silicon-nitride[50].

Preparation of alpha-silicon-nitride from $SiCl_4$ has also been reported by Kasai and Tsukidate[51] of Tokyo Soda Manufacturing Co., Japan. In this process, $SiCl_4$ and ammonia are continuously reacted in an inert gas atmosphere at 10-30°C and the resultant $Si(NH)_2$ containing ammonium-chloride is decomposed to obtain alpha-silicon-nitride of small and homogeneous grain size. Thus 33 parts of $SiCl_4$ saturated in nitrogen at 25°C and ammonia (20 g/hour) were fed through an outer and inner tube of a double-cylinder into a reaction tube (diameter 60 mm and length 250 mm) water cooled at 10°C; the product was collected in a vessel under the tube and heated to and at 1400°C and 600°C/hour, for a period of 2 hour in nitrogen atmosphere. Alpha silicon nitride with particle size of 0.1-3 μm (diameter) was obtained by this process with purity above 99.9 %.

Perugini[52] at Inst G Donegani, 28100 Novara, Italy, produced silicon nitride from the same starting materials ($SiCl_4$ and ammonia), using an arc plasma furnace. The product was amorphous silicon-nitride with a particle size of 50-150 Å, which was later transformed to preferred alpha structure by heat treatment in nitrogen atmosphere. Faculty of Engineering, University of Tokyo, Japan has reported using a radio frequency plasma torch combined with a DC plasma torch for generating stable and enlarged argon plasma, to which a mixture of $SiCl_4$ and ammonia was introduced. The reaction product deposited on the inner wall of the water cooled copper coloumn or mould, was in the form of very fine powder of silicon nitride.

Purification and estimation of trichlorosilane (TCS)

In literature purification of TCS by chronatographic method using activated alumina, activated silica gel on ferrohydroxide, as well as by complexation of impurities with nitrobenzene, triphenylchloromethane, acetonitrile, etc. has been mentioned. But in practice fractional distillation in a highly efficient distilling column seems to be the most economical and commercially viable route. Distillation of the usual mixture of TCS with $SiCl_4$ formed in commercial TCS production units, delivers very good separation of the above two compounds, as their boiling points are well separated. Nevertheless, design of an highly efficient distilling apparatus with large number of theoretical plate offers the solution of getting high purity TCS from its mix.

Boron and phosphorus are the two most worrisome contaminants in TCS, as they are electrically active. Accordingly their concentration in the final product needs to be brought down to ppm (or preferably ppb) level. To remove these two undesired elements, the B(III) and P(III) chlorides are first converted into less volatile compounds by complexation reaction; the mixture then put into fractional distillation to separate out TCS. This procedure is capable of bringing down concentration of both boron and phosphorus to 1 ppb level. One example of complexation reaction would be, reaction with p-hydroxyazobenzene or phenylazonaphthol, which forms strong complexes with boron halide.

TCS content in the distilled product can be ascertained by contacting the solution with a solution of sodium-hydroxide, producing hydrogen gas, which is measured accurately under NTP condition to determine TCS content:

$$HSiCl_3 + 4\ NaOH = H_2 + Na_2SiO_3 + 2\ NaCl + H_2O + HCl$$

The measured volume of hydrogen by the above process is directly proportional to TCS content in the mixture. TCS can also be determined by gas-chromatography using a special column of Silicone DC-550 at 15 wt% Celite of 80-100 mesh as stationary phase at 30°C and carrier gas N_2 and its flow rate at 19.3 cc/min. The accuracy of this method is quiet high. Very small quantities of other impurities in TCS (of the order of 10^{-4} to 10^{-6}%) can only be determined by generating silicon rod by thermal cracking TCS, and measuring electrical resistivity of the above silicon rod.

Thermal cracking of trichlorosilane (TCS)

TCS can be thermally cracked on a filament rod up to a diameter of about 30 mm of polycrystalline silicon. The thermal cracking is carried out in presence of hydrogen and the process takes place during progress of silicon deposition:

$$4\ SiHCl_3 + H_2 \rightarrow Si + 3\ SiCl_4 + 3\ H_2 \quad \text{(endothermic)}$$

The other reaction, shown below, do not take place easily:

$$SiHCl_3 + H_2 \rightarrow Si + 3\ HCl \ \text{(endothermic)}$$

The thermal cracking as per above reaction, takes place at about 1200°C, generated by a high amperage and low voltage current to the heating filament. Since silicon has high resistance (of the order of 100 ohm.cm) at room temperature, it can be easily heated with our regular voltage supply (220 V) low current in the initial phase of heating; but since it has a negative value of specific resistance, it becomes conducting at around 350-400°C; and thus both high tension and low tension electric circuit is used in heating silicon rod in a thermal cracking unit. Both these circuits are provided with saturable reactors, which possess more reactance than resistivity of load, so as to stabilize the electric current. The characteristics of the silicon formed varies with temperature; for example, at the initial stage around 700-800°C temperature, when silicon first starts depositing on the filament, it looks like a brown powder (amorphous silicon) and with rise in temperature, the deposit takes a compact metallic structure and looks glossy white.

Besides TCS (whose purification procedure was mentioned earlier), the other gaseous component (hydrogen) used in the cracking process also needs to be highly purified to get the desired purity in the deposited silicon. Hydrogen purification is done by passing the gas through a vessel loaded with a saturated solution of potassium permanganate (made slightly alkaline) which retains volatile hydrides such as arsine, phosphine, etc. by their oxidation. Then the gas is first passed through calcium chloride and then through platinized asbestos kept at 400-450°C, which eliminates oxygen by catalytic burning with hydrogen. Moisture generated in the last step is then removed by passing the purified gas through two columns of silica-gel and the carbon-dioxide removed through the second column of soda-soaked asbestos. Finally the gas is passed through a column of glass marbles greased with silicon lubricants, in order to arrest aerosols. Hydrogen thus purified contains 0.1-0.2 ppm of oxygen. Following purification, the hydrogen stream is divided into two paths—one part draws the TCS into the cracking chamber and the other gas used as makeup hydrogen for the thermal cracking chamber.

Besides purity, molar ratio of hydrogen with respect to TCS plays a crucial role in the basic cracking phenomenon in the high temperature pyrolytic reactor. For example, if the molar ratio of TCS to hydrogen is less than 10, silicon is produced principally by the following reaction:

$$SiHCl_3 = \frac{1}{4} Si + 3/4\ SiCL_4 + \frac{1}{2} H_2$$

But when the above molar ratio exceeds 20, the following reaction predominates:

$$SiHCl_3 + H_2 = Si + 3\ HCl$$

Similar reactions have been reported by Naka *et al.* The reaction of TCS at high temperature is a complex one and at high temperature $SiCl_2$ is also reported to occur. At zero hydrogen pressure, the thermal decomposition at 800-900°C proceeds according to the first equation above, and at hydrogen pressure greater than the pressure of TCS, the high temperature reaction follows the second equation above. Thus optimum temperature for pyrolytic decomposition of silicon from the above mixtures is found to be around 1125°C. Yoshizawa reported thermal decomposition of TCS as per the following three routes:

$$4\ SiHCl_3 \rightarrow 3\ SiCl_4 + Si + 2\ H_2$$
$$2\ SiHCl_3 \rightarrow SiCl + Si + 2\ HCl$$
$$SiHCl_3 \rightarrow SiCl_2 + HCl$$

The decomposition process is affected by flow rate of the gas as well. The intermediate product, gaseous $SiCl_2$, explains the etching and dissolving of silicon, when hydrogen is in slight excess. Accordingly, TCS is fed at a definite ratio along with hydrogen and at fixed velocity input into the pyrolytic chamber. General efficiency of silicon deposition is around 30 %. If a single crystal silicon filament is used in the cracking chamber, the final product is also obtained as a single crystal. These single crystal silicon filaments (1-2 mm in diameter) are grown from silicon melt by dipping a small single crystal in contact with molten silicon, and continuously pulling the crystal away to form the filament. Niederkon *et al.* reported the highest yield by this process as 60.4% at 1500 K and at a hydrogen to TCS ratio of 49:1. The reduction of TCS at 1100°C by hydrogen gave highly pure silicon with 88.5% yield. At low hydrogen concentration, the decomposition takes place predominantly as per the following route:

$$4\ SiHCl_3 = Si + 3\ SiCl_4 + 2\ H_2$$

Simultaneously, the heart of the cracking system, the pyrolysis chamber, has to satisfy a number of design criteria, in order to achieve the desired silicon rod quality. Some of these essential criteria are–uniform deposition of silicon from the TCS mixture around the filament rod of silicon, which governs ultimate shape of the product rod, facility for easy manipulation and safety, etc. Accordingly, the cracking chamber is usually designed as a horizontal cylinder (especially with single filament deposition) or in the shape of a bell-jar for multiple rod deposition simultaneously. In both the cases, entry and exit of the gases are so designed, in order to envelope the filament rod uniformly all around. In multiple rod system, the gas entry point is usually at the center of the chamber and the seed rods are placed uniformly in a circular fashion around it. All parts of the installation that come in direct contact with the reaction mixture are made up of quartz, stainless steel, teflon or molybdenum. Especially all threaded parts are made with molybdenum, while dome or the cover is made from quartz for better visibility.

While the cracking is in progress, the quartz dome is kept cold, in order that the heat from silicon deposition is available only to the rod surface and simultaneously see-through quartz dome remains clean. Deposition rate in such thermal cracking unit is generally faster than a conventional CVD unit. Nevertheless, the former process produces polycrystalline variety of silicon while the later, directly grows single crystal, if a monocrystal slim rod is used. Polycrystalline silicon rod produced by the above thermal cracking of TCS, is later grown into a single crystal rod by floating zone refining machine, or single crystal ribbon cell by continuous pulling from its melt. CVD process of single crystal silicon making is expensive as the process is slow and also produces low quality single crystals. Theoretically, the highest growth rate possible by this process is 2 mm/hour, but usually commercial production rates are still slower (1 mm/hour or less). Attempts to grow silicon at a faster rate has resulted in defects such as dendritic growth, which produces rough surface, voids and cracks. These defective feed rods would either crack or emit silicon particles on melting during subsequent float zone growth. Recently, Keck *et al.*[53] at Advanced Silicon Materials, USA, described a CVD method for producing semiconductor grade polycrystalline silicon rod with a larger diameter from silane and halosilanes, in which an AC current was provided to the seed rod by a power source with fixed or variable high frequency in the range 2-800 kHz to produce a skin effect, that concentrates more than 70 % of the current in an annular region comprising 15 % of the rod volume.The entire volume of the rod was maintained at a temperature of 600-1200°C with a temperature variation of less than or equal to 50°C. The polycrystalline silicon grown by this process, has a diameter greater than 150 mm.

Production of semiconductor grade silicon from silane in a fixed bed has been described by Niedam *et al.*[54], and production data as well as reactor configuration are shown in Table 7 below.

Table 7. Production of semiconductor grade silicon from silane[54].

Reactor size (mm)	100 (bottom)	70 (top)	700 (high)	
Diameter of reactor tube (mm)	30	50	70	100
Optimum flow of HCl/hour (in litre)	20	50	80	120
Optimum flow rate (liter/h cm^2)	2.8	2.5	2.1	1.5
Per unit cross-section flow (cm/sec)	0.78	0.695	0.585	0.415
Column packing	1.37 kg/dm^3 (with 3 mm granular silicon).			

Beckloff *et al.*[55] reported CVD deposition of pure silicon from $SiCl_4$ and hydrogen mixture with grain size as large as 15-20 μm for a coating thickness of 50 μm on TiB_2 substrate.

CVD process of silicon crystal growth being a slow process, much attention has been paid for innovations to accelerate it. One significant achievement in this direction was the advent of plasma enhanced CVD (in short called PECVD)[56]. Torres *et al.* from Inst. of Microtech, University of Nenchatel, Switzerland, recently[57] described one such process, whereby microcrystalline silicon solar cell having conversion efficiency above 5% was deposited at a deposition rate in excess of 10 Å/sec. This was achieved by a VHF-GD instrument generating excitation frequency of 130 MHz, by increasing the plasma power at dilution ratio of 7.5 % of silane in the silane/(silane + hydrogen) ratio. The investigators also observed a morphological transition in the process, as it advances from amorphous silicon to microcrystalline silicon in the initial phase of deposition, which was followed by a rapid increase in deposition rate. This indicates that crystallographic nature and ultimate efficiency of the silicon cells manufactured by this process vary in a significant way depending on the process parameters set during its manufacture. Rapid thermal CVD has also been reported from China in recent years, by Zhao *et al.*[58], who reported growing polycrystalline silicon thin film (10-20 μm in thickness) from SiH_2Cl_2 or $SiCl_4$ by rapid thermal CVD technique, with a growth rate up to 100 Å/sec at the substrate temperature T_s of 1030°C. Average grain size and carrier mobility of these films was found to depend upon T_s. The films were grown on heavily phosphorus doped silicon wafer for solar cells and energy conversion up to 9.88 % obtained at AM 1.5G (100 mW/cm^2) at 25°C. Faster deposition of silicon crystals fromTCS using PECVD has also been reported by Rostalsky *et al.*[59] (Department of Semiconductor Technology, Technical University of Hamburg, Germany), who carried out the above mode of deposition at a lower temperature on graphite substrate. Here the investigators paid special attention to growth of larger grains based on optimization of process parameters. Chlorine content in such silicon deposits were also estimated by energy dispersed X-ray technique. Ito *et al.*[60], on the other hand, (at Toyota Central R&D Laboratory, Aichi, Japan) paid attention to the effect of hydrogen on silicon deposition by PECVD method. He established through his experiments that a film which contains more hydrogen atom is more amorphous, and there are fewer hydrogen atoms (less than 1 at %) at the film/substrate inter-phase in the polycrystalline silicon films. They inferred that dehydrogenation process precedes crystallization in such a deposition process. Dehydrogenation step induces redistribution of silicon atoms and thereby facilitates crystal formation.

Dairiki *et al.*[61] at Tokyo Institute of Technology (Department of Electrical and Electronics Engineering, Tokyo, Japan), studied the effect of SiH_2Cl_2 concentration in mercury sensitized CVD process. In their study SiH_2Cl_2 and hydrogen was used for preparation of amorphous silicon in the CVD apparatus. SIMS measurements in these experiments indicated that chlorine was incorporated into the deposits of silicon film at a concentration up to 10^{18} atom/cc, even at a flow rate of low hydrogen concentration (SiH_2Cl_2:H_2 ratio of 0.01). Excess chlorine incorporated in the deposit was subsequently removed by hydrogen with high dilution. Excess acceptor concentration was thus controlled by SiH_2Cl_2 input into the CVD apparatus. These techniques resulted in increase in quantum-efficiency of i-layer, which is especially prone to light-induced degradation. Thus a highly stabilized 9 % efficient (1 cm^2 single junction) amorphous silicon solar cells were produced by this technology.

Guo *et al.*[62] at Electro-Technical Laboratory (Thin Film Silicon Solar Cell Super Laboratory, Ibaraki, Japan), reported high deposition rate of silicon (hydrogenated micro-crystalline) at a relatively high working pressure in a PECVD apparatus. Their finding was that deposition rate of silicon exhibits a maximum at about 4 torr pressure and crystallinity of the film decreases linearly with increasing pressure. Thus control of both silane depletion rate (precursor used in these experiments) and high pressure applied in the plasma, is necessary to improve the crystallinity of silicon deposited at high rate by this process. Silicon deposition rate reported in this work was 9.3 Å/sec.

FLUIDIZED-BED THERMAL CRACKING:

Breneman *et al.*[63] demonstrated in 1978 that cracking of silane in a fluidized-bed of silicon particles could generate solar grade silicon at a cost less than $10/kg. He used metallurgical grade silicon as a starting material to produce TCS by reacting it with $SiCl_4$ in the presence of copper catalyst. The TCS thus produced was further converted to silane by disproportionation reaction, which was later thermally cracked in a fluidized bed reactor to produce high purity silicon.

Ethyl Corporation (Richmond, VA, USA) in their 1250 metric ton/year polycrystalline silicon plant at Houston, Texas (started in August 1988) used fluidized-bed thermal cracking unit with tiny silicon particles with silane/hydrogen mixture, to produce spheres averaging 700 µm in diameter[64]. They also claim that the plant in total uses only one-tenth the energy of the best Siemens process. Because of the greater surface area, fluidized-bed reactors are much more efficient than filament or rod reactors mentioned above. Texas Instruments, USA has operated its own, proprietary fluidized-bed plant since 1978 using TCS. But one advantage is that, fluidized-beds are at dynamic equilibrium condition throughout its operation, while filament or rod reactors, mentioned earlier, are in unsteady state. Thus one fluidized-bed reactor can do the work of 20-50 rod reactors in the same time space.

J.C. Schumaker Co, Oceanside, CA (USA) is developing a process that combines tribromosilane with fluidized-bed deposition. It is a closed loop process without any waste or environmental degradation. Here metallurgical grade silicon is reacted with hydrogen and silicon-tetrabromide to form a mixture of silicon-tetrabromide and tribromosilane ($SiHBr_3$). The later decomposes at about 800°C to deposit pure silicon on fluidized silicon seed particles, while the rest of the gas is recycled to the first reactor. This process requires about one-fifth the energy and half the capital cost of the Siemens process, and uses about 70 % less labour charge. The process has been tested in a 1 ton/year pilot plant and currently testing a 40 ton/year plant for commercial use. Great Lakes Chemical Corporation, West-Lafayette, Indiana (USA), is also working on a bromine based plant.

Recently, Schreider and Kim[65] at Wecker-Chemie GmbH (Germany), patented a fluidized bed process, which produces high purity granulated silicon with chlorine content less than 50 wt. ppm. The reactor in this case was divided into two zones to facilitate both heating and reaction separately. The particles are fluidized with an inert silicon-free carrier gas and heated by microwave energy and through which a reactive silicon containing gas is injected. The average temperature of the reaction zone is maintained below 900°C, and that of the fluidized silicon particle zone was maintained above 900°C. The silicon containing gas was introduced in the reaction zone through a tubular nozzle. Temperature of the reaction gases is controlled by the height of the fluidized-bed and the minimum fluidization gas velocity. Suitable reacting gases like $SiCl_4$, TCS, and SiH_2Cl_2, as well as suitable carrier gases like hydrogen, nitrogen, argon or helium has been tested successfully. Silicon granules produced by this process is pure enough for manufacture of semiconductors and silicon solar cells.

Fluidized-bed of silicon particles has also been used to manufacture methyl-chloro silane. Rodrigue *et al.*[66] at Powder Metallurgy and Magnetic Material Laboratory (Switzerland), reported recently one such process, where methyl-chloride gas was used as the reacting as well as fluidizing gas in the fluidized-bed reactor filled with silicon particles and a catalyst. Yield of silicon from such a process was quite good when process parameters are optimized. High purity silicon can be produced by thermal cracking, in turn, from this methylchloro silane by atomization technique described by abov.e investigators.

Besides TCS and silicon-tetrachloride, other silicon compounds have also been used to generate high purity (solar grade) silicon. In the following few pages we will discuss these techniques as well as the present trend of research in this direction.

(b) Preparation of silicon from SiI_4

Preparation of silicon by the decomposition of SiI_4 is a straightforward process, and apparently because of the ease of purification, this process was adopted to produce extremely high purity silicon. In this method, low pressure is essential to have a reasonable deposition rate. Necessary steps for production of silicon by this process involves –(i) reaction of commercial grade (metallurgical) silicon with iodine, (ii) purification of resultant SiI_4 from mixture, (iii) distillation of the iodide to get pure product, (iv) decomposition of SiI_4 to form silicon and iodine as by-product, and (v) iodine recovery for reuse. Economics of the process depends to a large extent on recovery and reuse of iodine. Another more elaborate process involves an additional purification step-zone refining of SiI_4 before distillation. SiI_4 is decomposed at a temperature higher than 1200°C on the hot surface of pure silicon filament. SiI_4 can be recrystallized with toluene, sublimed under reduced pressure, rectified in an inert atmosphere, and purified by zone melting.

Rubin *et al.*[67] reported preparation of silicon by iodide process involving the following steps—reduction of silica to silicon, reaction of silicon with iodine to form SiI_4, distillation of SiI_4, sublimation, and finally thermal decomposition. Results obtained by the above investigators are shown in Table 8 and in temperature range 500-900°C.

In order to attain maximum productivity, the temperature of the iodine source, i.e. the iodine concentration in the argon flow, should be increased to a maximum.

It is well known that removal of boron along with other non-metallic elements such as phosphorus and arsenic, is generally very difficult. This is, however, easier in the case of SiI_4 as it has a boiling point sufficiently different from most of the other iodides. For more advanced purification, zone melting is carried out, which takes advantage of the different seggregation coefficient of impurities. A simple but efficient method of removing non-metallic impurities like boron, arsenic, and phosphorus involves fractional distillation in a quartz tube at atmospheric pressure.

Table 8. Yield dependence of SiI_4 at various temperatures[67].

Reaction conditions		Values of the parameters	
Amount of iodine per batch		100 g	
Amount of silicon		230 g	
Height of silicon column		20 cm	
Argon flow capacity		15 litre/hour	
Temperature of iodine source		190°C	
Duration of experimental runs		2 hour	
Sample no.	**Reaction temperature(°C)**	**Si content in SiI_4(%)**	**Yield(%)**
1	500	4.74	90.6
2	600	5.02	95.0
3	700	5.16	98.8
4	800	5.02	95.0
5	850	4.92	94.1
6	900	4.59	88.1

Silicon-tetraiodide has a melting point of 121°C and can be produced with a 70 % yield by passing a carrier gas saturated with iodine vapour over silicon containing 4 % copper[129] at 600-700°C. Silicon-tetraiodide can be prepared on kilogram scale in quantitative yield by passing nitrogen gas saturated with iodine vapour at about 200°C over granulated silicon at 1150-1200°C. Approximately 1.5 kg silicon-tetraiodide can be produced in 5 hour by this method[130]. Silicon-tetraiodide is also used for making silicon solar film by electrolyzing a solution of SiI_4 in aprotic organic solvent[131]

(c) Preparation of silicon from SiO

Recently, Kendo *et al.*[68] at Nippon Steel (Japan), reported a process whereby high purity silicon can be prepared by heating solid SiO at 1000-1730°C to convert the solid SiO into liquid or solid silicon as well as solid silica, through disproportionation reaction and spreading silicon formed from the SiO_2 and/or SiO. Above investigators reported manufacturing SiO by heating a mixture of carbon, metallurgical silicon and/or ferrosilicon with silica, to form a gas containing gaseous SiO and subsequently collecting SiO by cooling the gas.

(d) Preparation of silicon from sodium-fluorosilicate

Reaction between SiF_4 and sodium metal to produce high purity silicon has already been described in Sec 2.2. SiF_4 on the other hand can be obtained from phosphoric acid plant, which produces sodium-fluorosilicate as a by-product. At Paradip Phosphate, Orissa, this material causes environmental pollution and can be gainfully employed to produce silicon by the above process. Dr Angel Sanjurjo at Stanford Research Institute, Menlo Park (USA) earlier reported a one-step process for producing silicon from sodium-fluorosilicate, which costs only $5.0/kg (almost one-twelfth the current price at that time). The chemical reaction involved in the process, uniquely produces its own heat and separates silicon from other impurities in 'self-sustaining method'. Here the spontaneous reaction between sodium-fluorosilicate and sodium is used to obtain SiF_4, which in turn reacts with sodium to produce silicon-metal and sodium-fluoride.

SiF_4 has also been reported to form into silicon-fluorine-hydrogen amorphous alloy by exposing a substrate to the mixture of SiF_4 and hydrogen under glow discharge[69]. The thin film of Si-F-H alloy deposits on the substrate by this process. This material is claimed to be superior in performance to other amorphous solar cell materials (having efficiency of around 10 %). A small amount of this alloy is needed to fabricate the solar cell (typical thickness being 1 μm), and mass production is easier than pure silicon single crystal. This material also has a higher doping efficiency (i.e. a given amount of dopant will go a long way in boosting the materials conductivity further). Other advantages of this alloy are that, no structural changes are caused by sunlight and it does not lose hydrogen when exposed to solar heat. It is scratch proof and unlike some crystalline material, this substance is sensitive to the most intense portion of the solar spectrum. It has been claimed that solar devices can be fabricated from this material with electricity costing 50 cents/watt as against $10/watt for crystalline silicon (1980 cost level). These cells, known as 'ovonic amorphous silicon' semiconductor cell, was first reported by Stanford R. Ovshinsky, President, Energy-Conversion Devices, Troy, Michigan (USA) in 1980's.

(e) Preparation of silicon from SiO_2 by aluminium and magnesium

Alumina has a higher free energy of formation than silica, and therefore aluminium is capable of reducing silica. At a high enough temperature, if oxygen is not available in the atmosphere to combine with the aluminium, this reaction is likely to take place. But care should be taken to avoid contamination with other metal impurities, and only very pure and stable refractories such as fused cast alumina, magnesia, zircon, or beryllia should be used in contact with the exothermic reaction, producing high temperature. Similar is the case with magnesium and the chemical reactions involved in the process are shown below:

$$3\,SiO_2 + 4\,Al = 2\,Al_2O_3 + 3\,Si$$
$$SiO_2 + 2\,Mg = 2\,MgO + Si$$

(f) Preparation of high purity silicon by vacuum refining and directional ingot solidification

A group of workers at Kawasaki Steel Corporation (Japan)[70-72] has recently studied such a process, which is claimed to generate solar grade silicon.

Both plasma and electron-beam was used in melting and vacuum refining of the silicon metal, followed by directional solidification in a special mould. Phosphorus, boron, carbon, oxygen was vacuum refined by this technique and iron, aluminium, titanium and calcium removed during directional solidification. Both the operations were carried out in a special apparatus, thus saving energy as well as time of processing, and turning out solar grade silicon at a lower cost. The apparatus has the facility to irradiate the melt surface and also cooling facility at the bottom for creating temperature gradient and consequent directional solidification. Thus the volatile impurities were removed from the surface of the melt by applying vacuum simultaneously, while the melt over flow into the casting mould for removing other impurities by directional solidification[73]. In the plasma process[72] following vacuum refining, the melt was treated with plasma of water and hydrogen in an inert gas to remove boron and carbon together with iron, aluminium, titanium, calcium and oxygen. The molten silicon may also be treated with plasma before vacuum refining. Hanzawa *et al.*[74,75] melted the crude silicon in a special graphite container by electron beam, and the graphite container was coated with silicon-carbide or carbon releasing agent, which besides facilitating release of the cast also protected the crucible and enhanced its life considerably. Stirring during electron beam melting of silicon, is also reported to improve purified silicon yield[76,77]. Besides controlling cooling from bottom of crucible/mould, inert gas is also blown above solidifying boundary, which further facilitates the purification process[78]. In these studies, above investigators also successfully purified silicon to solar grade, by pouring molten silicon dropwise down through a furnace at less than or equal to a pressure of 1×10^{-2} Torr, in order to increase surface contact and better refining of silicon through complete removal of phosphorus, aluminium, and calcium by vaporization[79]. In yet another novel attempt[80] above, investigators removed boron by a portable power source to generate plasma-arc and spraying molten silicon in an oxidizing gas through the arc; simultaneously external magnetic field was applied perpendicular to the plasma axis, at the rapidly moving contact point between the silicon and plasma. During purification of silicon by directional solidification process, the solidification rate was determined by monitoring the position of the solidifying boundary[81]. It is necessary to continuously adjust the thermal gradient in the mould during the course of solidification in order to achieve desired purity. This control is generally achieved through placing a heater above the mould and a cooling arrangement at the bottom. Temperature sensors are placed at various heights in the mould wall, or an ultrasonic distance detector placed above the mould, in order to assess solidifying rate from the recorder temperature or ultrasonic apparatus output. These procedures are being widely adopted in implementing this technology for controlling cooling rate at a predetermined rate. A single crystal substrate is placed at the bottom of the crucible before pouring molten silicon into it for unidirectional solidification[82].

(g) Preparation of high purity silicon by sulphide reduction process

Tamura[83] at Nippon Steel, Japan, reported a process in which silica along with sulphur or sulphides is mixed and fused, followed by reduction at a temperature equal to or below the melting point of silicon with the addition of reducing agents. The result is the formation of high purity silicon suitable for substrate formation and manufacturing semiconductor devices as well as solar cells. Silica in this process may further be added along with the reducing agent as necessary.

(h) Wet method of producing high purity silicon powder

High purity silicon powder having uniform particle size is produced[84] by crushing chunk silicon or cutting scraps from silicon wafer manufacturing plant (a waste), and treating the mass with a solution of H_2SiF_6, alkali metal hydroxide, or ammonia. Alternatively, the silicon mass is treated with aquous alkali-metal hydroxide and then treated with aquous H_2SiF_6 to precipitate out pure silicon. The aqueous solution may be diluted with metal-ion free surfactants to ease the process. Silicon thus produced is suitable for solar cell manufacture and use in ceramic industry. Another aquous method involves mixing metallic silicon powder having maximum grain size equal to or greater than 30 μm with aquous binder solution, granulating and then heating to 1150-1350°C under inert atmosphere containing 1-10 vo1% of methane. The process generates silicon granules with specific surface area greater than or equal to 0.3 m^2/g, oxygen content less than 0.5 wt% and porosity 35-70%. The same process can be extended to manufacture silicon-nitride by nitriding the above prepared granules of silicon at 1150-1400°C in nitrogen atmosphere, containing ammonia or hydrogen. The product is an alpha-type silicon-nitride with good homogeneity and uniformity.

Solar grade high purity silicon has also been generated[85] by hydrolysis of purified tetra-alkoxy silane containing boron less than 0.1 ppm and phosphorus less than 0.1 ppm, as well as total impurity below 20 ppm. Purified tetra-alkoxy silane for this purpose was, in-turn, prepared from precision distillation of commercial grade tetra-alkoxy silane (e.g. tetramethoxy silane).

(i) Fused salt electrolysis for producing pure silicon

At NREL (Golden Co, USA) Moore *et al.*[86] electro-deposited semiconductor grade silicon from crude molten silicon bath. The crystalline silicon thus produced has thickness of 5-50 μm and was deposited on heavily doped phosphorus (III) type Czochralski silicon and polished silver. Growth of crystalline silicon from metallurgical grade silicon bath was demonstrated for the first time. A 50/50 mixture of KF and LiF was used as solvent and K_2SiF_6 as solute in the process. The films were obtained with relatively low impurity, thus demonstrating self-purification effect of electro-deposition. Films were grown at temperatures ranging from 750 to 850°C, and the best film quality as well as deposition rate was recorded at 850°C. The film quality was also improved when a dissolving silicon anode was used to replenish the silicon concentration in the solution. Deposition rate on to (III) silicon was higher than those previously reported and were achieved through use of very low resistivity (0.013 ohm-cm) silicon electrodes. The films were deposited at a current density of 10-15 mA/cm^2 and as grown layers were n-type.

Ginatta[87] at Cathingot Ltd, Liechtewstein, also described a similar electrolysis mode for electrolytic deposition of high purity silicon from molten metallurgical grade silicon bath.

(j) Production of silane by chemical reactions in molten salts

Silanes of commercial interest as starting material for producing semiconductor grade silicon, or for hydrosilylation reactions Me_nSiH_{4-n} (n = 0-3) have been prepared[128] by direct hydrogenation of the corresponding chlorosilanes in the presence of various interstitial hydrides of transition metals, preferably of titanium, which are generated *in-situ* in chloroaluminate melts and with aluminium as halogen acceptor.

2.3 ANALYSIS OF THE PURITY OF SILICON

In the wet method of analysis, silicon is first converted to silica (SiO_2) or silicic acid by fluxing with oxidizing agents (like sodium-carbonate or sodium bromate $Na_2Br_4O_7$), followed by acid precipitation. Its yellow compound (silico-molybdate) can be estimated colorimetrically.

Among instrumental methods or analysis, it can be analyzed spectrophotometrically, as it gives two intense lines at 251.6 and 288.2 nm. Other instrumental methods of analysis include:

(a) Neutron activation analysis
(b) Scintillation spectrometry
(c) Mass spectrometry
(d) Emission spectrography

Among the above four methods, the 'neutron activation analysis' is used most and we will discuss the method in a greater detail.

NEUTRON ACTIVATION ANALYSIS OF SILICON:

In this method, the sample is bombarded by thermal neutrons, so that impurities are converted to radio-isotopes. With both silicon and germanium, the host elements are also rendered radioactive. Therefore, each impurity is separated chemically after adding a known amount of the inactive impurity to serve as carrier. The percentage recovery of the impurity is determined, together with its activity. Allowing for the material not recovered, the radioactivity is directly proportional to the original amount of the impurity and to the condition of activation. The latter are not easily determined. So a comparative method is adopted, in which a standard of known impurity content is irradiated at the same time as the analysis sample. In semiconductor analysis, the method was first used to determine arsenic and germanium. With silicon, activation analysis has been used to determine a wider range of impurities than in germanium. The sensitivity range by this method varies between 10^{-2} and 10^{-5} ppm.

CHAPTER 3

Crystallization of Silicon

3.1 PRODUCTION OF SILICON SINGLE CRYSTAL FROM POLYCRYSTALLINE SILICON

Method of production of polycrystalline silicon has been described in the previous chapter. To generate silicon single crystals from the polycrystalline silicon the silicon mass is remelted and slowly grown against a single crystal seed. This method of production of single crystal is necessary if the silicon is particularly to be used for fabricating semiconductor devices. This process of single crystal growth not only gives maximum degree of crystal perfection in large size by removing grain boundaries, lattice defects (dislocation) and thereby increasing carrier mobility, but also extends another opportunity to remove minutest impurities, thus generating hyperpure silicon. Various methods have been adopted by various investigators to meet this technological perfection, and we will discuss them in short, in the following few pages.

(a) Zone refining technique for generating silicon single crystals

Commercially there are two distinct approaches to zone refine polycrystalline silicon. These are float-zone (FZ) and Czocharlski method (CZ). In the former method the melt zone created on the silicon rod by sharp narrow heating is moved from one end to the other for taking out the impurities continuously from one phase and concentrating at the other. In Czocharlski method, the crystal is pulled from the bath and sometimes the crucible is also rotated to facilitate the process. FZ method nowadays produce about 15 % of the commercially available silicon single crystal, while CZ method the rest 85 %. As much as 40 % of the commercial photovoltaic-cell and micromechanical devices available in the market use single crystal silicon. The CZ-crystals are used mainly for manufacturing highly integrated low-power devices like microprocessors, RAM's, DRAMS's, ASIAC's, etc, whereas FZ-crystals of silicon are used mostly in making discrete or low-integrated high-power devices such as diodes, transistors, and thyristors. Although attempts have been made to purify silicon rod in horizontal fashion in FZ technique, the problem faced are in a horizontal furnace in FZ technique, using either very thin quartz boat or with induction heating a water cooled container (solidified silicon adheres strongly to quartz, so that thick boats crack but with water cooling there is neither any sticking nor any reaction); nevertheless commercially accepted technique involves vertical placement of the rod without any contact with the container, and moving a narrow melted (25-30 mm) zone upward. The heating is done by a microwave heater and impurities accumulate and finally come out at the other end of the rod, which is cut and discarded later. In this type of vertical zone refining, the process needs several passes to achieve the desired purity in the silicon rod. FZ-silicon growth rate is about twice as that of CZ-silicon. Morever, FZ-method has the advantage that the silicon rod do not come in contact with any container (silicon melt is very reactive in nature), whereas CZ-method needs some container to hold the melt; accordingly FZ-crystals are more pure than CZ-crystals. Typically, CZ-grown crystals contain 10^{16} atom/cc of carbon and up to 10^{18} atom/cc of oxygen, whereas conventional FZ-crystals contain two orders of magnitude less of these impurities. FZ-crystals can give up to 100 k.ohm.cm resistivity with corresponding boron dopant content of about 1×10^{11} atom/cc. Because of high reactivity of silicon melt, choice of container for CZ method is restricted to pure quartz, but it is also attacked to some extent by the melt, the silicon picking up oxygen, boron, and possibly other impurities from the container. Nevertheless, technically, FZ method is much more complex than CZ method. Commercial FZ-crystals are now available with lengths above 1 metre and diameter over 10 cm. Since the molten zone floats against the gravity, considerable skill on part of the technicians is needed to operate FZ-machines. These machines require induction heating at a frequency of 1-3 MHz, whereas CZ-method needs power of the order of 150 kVA, with crucible diameter of 30-40 cm, and silicon[88] charge 20-60 kg. The silicon that is put in CZ-equipment is either in the form of broken rods or silicon beads or granules. The latter form of silicon has advantage over broken rod from the viewpoint, that equipment can be run on continuous mode rather than batch mode used for broken rods. Besides increasing production efficiency, this helps to improve crystal quality and to reduce the size and the cost of the equipment, as the industry scales it up to larger crystals. In CZ method crystal growth is initiated by dipping the tip of a seed single crystal at the surface of the melt, and pulling the crystal as it grows. This pulling rate is about 2.5 mm/min, and crystal lengths of 1 meter and above with diameter up to 15 cm are now attainable, with the present commercial skill. Ferrofluid Corporation (Nashua, NH, USA), the leading maker of crystal growing unit, has already commercialized continuous crystall puller with continuous feed of fine granules into a steady state melt. FZ-method produces single-crystal silicon rod with lower oxygen content than CZ-method, but again the cost is high and used only by a handfull of commercial manufacturers round the world, like Wacker-Chemetronics, Germany and Shin-Etsu Handotal Co. of Japan. Wacker-Chemie grows up to 8 inch FZ-crystal of silicon, while Shin-Etsu 6 inch diameter rods. It may be noted here that, directional solidification in mould mentioned earlier generally results in multi-crystalline ingot, because of grain nucleation at the container wall[89,90]. A vertical gradient freeze method was developed at the AT & T Engineering Research Center (Princeton, NJ, USA) and is being used at AT & T's Reading, PA (USA) plant. Here the charge is put in a crucible that has a seed at the bottom and whose walls have resistance heater that provide a thermal gradient from top to bottom for directional ingot solidificaton. AT & T is producing 2 inch diameter silicon crystals by this method. A great advantage of such process is that, temperature gradient and growth rate can be adjusted to any desired value, and cooling rate can be controlled to minimize dislocation. A modified Bridgman process has recently been commercialized by Crystal Specialities Inc. (Colorado Spring, Co, USA), who is growing rectangular shaped crystals, from which 2 inch diameter wafers are cut. Cost of these machines, producing high quality crystals, are about $100,000 – $120,000. M/A-COM in USA has also developed an electrodynamic gradient furnace that is a combination of Bridgman and gradient freeze (made by Mellen Co, Penacook, NH, USA) having 20 electronically adjustable heat zone.

If the end product of either crystal growth or refined silicon crystal is in the form of a rod, it must be cut (e.g. by a special wire cutter, or saw) in order to make usable cells. This cutting process incurs a loss of about 30-40 % of the costly material, in the form of non-usable silicon dust. To avoid this loss, a new technology called ribbon or dendrite growth of silicon crystal has been developed.

(b) Growing ribbon or dendritic silicon crystals

Ribbon cells are drawn from a tank of molten silicon through a slot of graphite plate. Westinghouse Corporation (Pittsburg, PA, USA) commercialized this silicon sheet making technology which resulted in the development of economic solar cells. Again this method of growing silicon sheet is divided into three categories depending upon the three different types of meniscus used in pulling the film. This is shown in Fig. 6. Attempts have also been made to grow silicon from solution of other metals as well using this technology. In the later case, solubility of crystals in the metal solvent is such that the product is a dilute alloy. One such example is the growth of silicon crystal from solutions of aluminium and gold.

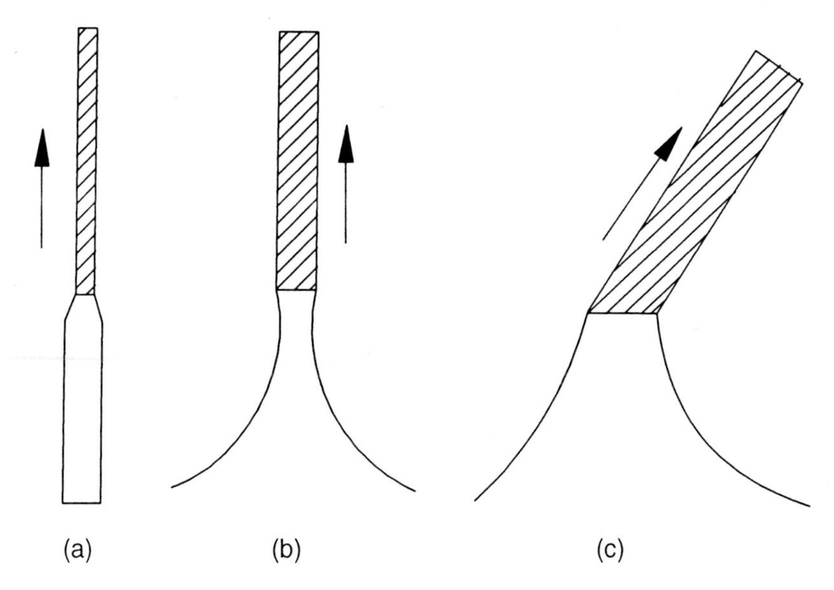

Fig. 6: Three types of meniscus formed during pulling of silicon film.

Growth of thin film crystals from vapour phase (Chemical Vapour Deposition or CVD) has been discussed earlier, which along with the ribbon drawing and dendritic growth are the most potential tool proven for commercialization, which has been tested in the last one decade.

Dendrites are the tree-like growth observed when silicon metal crystallizes from supercooled melt. The Westinghouse process[91] for making dendritic silicon photocell involves, pulling the dendrite web (a ribbon form of the single crystal silicon) looking like a very thin mirror strip of silicon, up to 4 cm wide and drawn from a bath of ultra-pure molten silicon at the rate of several centimeter per minute. It is wound upon a large diameter reel before being cut up into 10 cm lengths. Then the p-n junctions are created by diffusion, the cells are dip coated with an anti-reflective layer, 'windows' are opened by photolithography and finally the metal contacts are electroplated on to it. The result is the cells which operate around 15 % efficiency and which can be arranged on a 1 square foot module with packing factor of 97 %.

(c) Epitaxial growth of silicon single crystals

Epitaxial or oriented growth of silicon crystals on substrate has been practised commercially for its economic advantage. For example, Dr. R. Thomas at Carleton University, worked on a method of producing pure silicon at low temperature from inexpensive grade of silicon, where inversion layer cells eliminated the need for high temperature furnace. Its thin upper semiconductor layer is created by simply allowing a drop of silicon dioxide to spread over the surface of a crystal wafer spinning in vacuum. A thin layer of oriented crystalline silicon was also grown on the surface of amorphous substrate by Dr. Smith at MIT-Lincoln Laboratory (Lexington, MASS, USA). The process called 'graphoepitaxy', possess the speciality that oriented crystal of silicon can be grown on a substrate which itself is not a crystal. It uses photolithography to etch a grating on to the surface of a silicon glass plate. A thin film of noncrystalline silicon was then deposited on this substrate and heated to near its melting point by scanning with a laser beam. When the silicon layer cools, it crystallizes and after several passes of the laser, becomes essentially a single crystal, uniformly oriented along the direction of grating grooves. It may be possible in future to use a moulding process to mass produce large sheets having the surface gratings required to induce oriented crystal growth by this technique. If the surface is not etched, the process yields many microscopic crystals with a variety of orientations. Moulding is a low energy process that has the potentiality of yielding large area solar cell at a relatively lower cost.

This multilayer technology has also been used to create p-n junction in silicon solar cell at a much reduced cost. One such process, reported by Dr. Martin Green from the School of Electrical Engineering, University of New South Wales, Australia (under research project from Australian research grant scheme), whereby a thin layer of oxide film was grown on silicon, and then a thin metal layer was created on this for both electrical contact to the cell and also introducing a rectifying junction in the cell (Fig. 7). Conventionally, creating a p-n junction (which is expensive and time consuming process) involves introducing a thin layer of impurities on the surface of silicon at high temperature. Accordingly, this new process cuts the cost of silicon solar cell production by one-third.

(d) Growing thin film silicon on insulator (SOI technique)

Growth of thin crystalline silicon film on various insulators (SOI technology) has recently been studied intensely for its commercial potential[92-95]. In

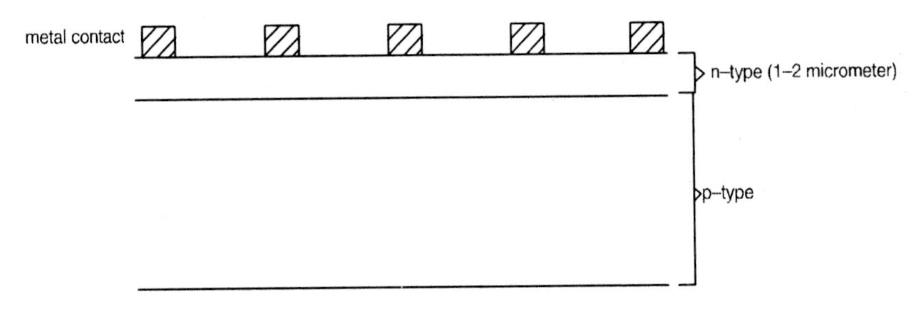

Fig.7: Typical p-n junction in a silicon solar cell

these studies, SOI structure is fabricated for creating electrically isolated complementary field effect (CMOS) transistors in top silicon layer. These devices are immune from hole-electron pairs produced in the bulk substrate by particles and other radiations. After its first reporting in 1980's, in subsequent years it opened the door to fabrication of thick film IC's, which in-turn resulted in much greater compaction of present day electronic devices, as this technology offers the opportunity of building two or more devices vertically up rather than conventional IC circuits spread out horizontally. In the choice of substrate for silicon deposition in thin film solar cell making by this technology requires a compromise between cost and quality[96]. Thus there are three generic types, namely–a transparent substrate (e.g. glass), an opaque substrate (e.g. a ceramic or metal), and a low-cost multi-crystalline substrate. Glass has the advantage of eliminating absorption within the substrate. However, the larger effective diffusion length, improved surface passivation, and increased process flexibility obtainable only with opaque substrate, particularly with low cost polycrystalline silicon, may considerably outweigh the modest optical benefit gained with a transparent substrate. The advantage in effective diffusion length that is required for a cell grown on an opaque substrate in order to offset the light trapping advantages of a glass substrate is by a factor of 2. Low pressure CVD with polycrystalline silicon has been used[97,98] for growing thin silicon film on a glass substrate.

In yet another method of microzone melting and recrystallization of silicon using a localized (20 μm to 3 mm) scanning heat source, Gat *et al.*[57] first showed that focused argon laser beam scanned over a poly-layer (unseeded) produced large grain size of silicon and showed dramatic improvement in carrier mobility. Production of single crystal SOI (Fig. 8) with seeded element (where the substrate is a silicon wafer which controls the therml expansion of the overall structure during heating, thus avoiding limitation of the problems like, silicon film cracking due to thermal expansion mismatch, was preferred in the commercial production of SOI.

Because of good electrical characteristics of the Si/SiO_2 interface, the lower substrate in SOI is generally 0.1-1.0 μm thick thermally grown SiO_2 (although silicon-nitride has also been used instead of SiO_2). The top polycrystalline silicon layer is 0.25-1 μm thick and chemical vapour deposited. Recently, Iwane *et al.*[99] described a process for producing a silicon semiconductor on insulator (SOI) substrate, which saves both material and cost in manufacturing silicon solar cell by these techniques. Here the substrate is separated by a porous silicon layer, which does not require a strong adhesion between the substrate and the single crystal grown. This sandwich mechanism creates a thin film with crystal on a low porosity and then a high porosity layer on the substrate. This separation is carried out by attaching a metal wire on the side surface of the substrate, through

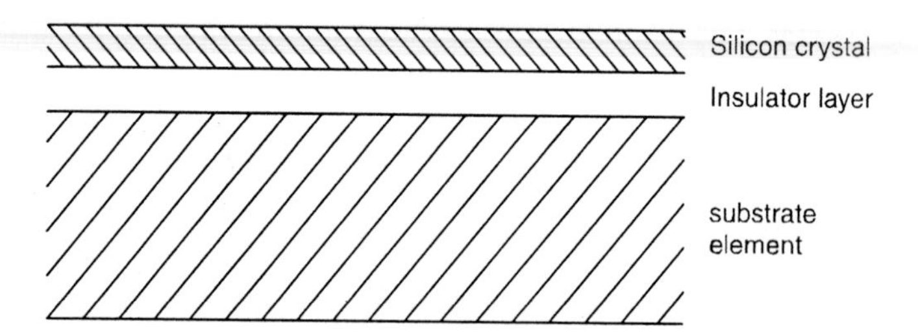

Fig.8: Schematic diagram of silicon on insulator (SOI) structure. (scanned crystal growth in above structure, for both seeded and unseeded form, is done by a heat scanning source on top layer).

which electrical current is passed, and the heat in the metal wire generated is transferred preferentially to the porous layer, thus performing the separation. The separated substrate is used for producing SOI substrate and the separated non-porous layer is reused in the process of producing SOI substrate. The heat source used in the microzone recrystallization is generally focused CW laser beam, argon, Nd/YAG and carbon dioxide lasers, or focused radiation from tungsten-halogen or mercury lamp, focused electron beam or unfocused radiation from graphite line heater. In practice production of whole silicon wafer SOI structure, use of line heater of various sources are preferred.

Polycrystalline silicon has also been deposited by rapid thermal CVD technique on silica, alumina, and mullite substrate by cold wall reactor, with high temperature reduction of TCS as precursor and boron-trichloride as dopant.[100] In this study (at Laboratorie PHASE, CNRS, Strusbourg, France) grain size and preferential orientation was found to be dependent upon deposition parameters as well as on the quality of the substrate.

Attempts have also been made[101] to deposit silicon on plastic film substrate at Fuji Electric Corporate R&D Center (Kanagawa, Japan). Plastic film substrate known as SCAF (series connection through aperture fomed on film) of size 40 cm × 80 cm was tested and the reliability towards light degradation of such cells were as high as that of glass substrate cell. Silicon crystals have also been grown on a cheap metal grade silicon substrate[102] by liquid phase growth using a metal solvent of indium, gallium, and tin. The process was successful in generating total impurity level as low as 10 ppm. The experiments were carried out in graphite, silicon-carbide, or silicon-nitride mould to form a plate shaped silicon layer. The method has added attraction of being able to bypass the need for a slicing step.

Recrystallized and epitaxially thickened polycrystalline silicon layer has also been grown on graphite substrate by electron beam zone melting process[103]. In this process, either vapour phase epitaxy or liquid phase epitaxy applied to enlarge the recrystallized seed layer to a thickness of 30-

40 μm. TEM studies reveal how epitaxy is influenced by silicon-carbide formed at the surface of seed layer. Efficiency of such solar cells have been reported[104] from 8.3 to 11%. Formation of carbide with graphite is an important parameter in such a case.

Attempts have also been made for faster deposition of silicon on foreign substrate by forcing the precursor gas at high speed through a jet nozzle pointed at the heated substrate and activating the gas by electron beam before it reaches substrate surface.[105] Efficiency of amorphous silicon cell generated by this technique showed solar cell efficiency of 8.7-9.4 % and deposition rate around 2-5 Å/sec. Cells formed by this method are quite stable.

Attempts have also been made to produce high purity silicon powder with uniform particle size, from molten silicon using atomizer (e.g. high pressure water atomizer) in a controlled environment.[106,107]

Polycrystalline silicon film (5-30 μm thick) has been deposited on glass substrate at relatively lower temperature (400-450°C) at the rate of 15 Å/sec by hot-wire CVD (in short called HWCVD) technique[108]. Undoped film formed by this technique has been found to have carrier concentration 10^{11}/cc and resistivity 10^5-10^6 ohm.cm, with activation energy 0.5 ± 0.05 eV and Hall mobility 14 ± 4 cm^2/V sec. By doping, resistivity can be varied by 6-7 orders of magnitude. Hydrogen passivation improves the product (carrier mobility, i.e. mobility lifetime improved) which is also carried out in the same reactor at a lower temperature (350-400°C). Later Bauer *et al.*[109] improved the HWCVD technique (by optimizing filament temperature, hydrogen dilution rate, gas pressure, etc.) for producing well stabilized amorphous silicon solar cell, with efficiency exceeding 10 %.

Besides affecting silicon deposition process itself by the above ingenuities, efficiency of silicon solar cells have also been enhanced by *in-situ* design modificaton of silicon solar cells. For example, there are two ways by which performance of present day silicon solar cells can be enhanced; these are – reducing thickness and grid shading. The ideal photovoltaic device will be 20-100 μm thick to satisfy the first criterion, while incorporating light trapping mechanism takes care of the second. Practicality also calls for these cells to be supported by a low-cost substrate. Aiken and Bernett[110] at University of Delware (Department of Electrical Engineering), USA, designed and tested one such thin substrate based crystalline silicon solar cell with no grid shading. In these designs contacts are sandwiched between a supportive silicon substrate and a 40 μm thick active silicon device layer. These cells have 535 mV (open-circuit voltage) and negligible shunt conduction and series resistance. Ford *et al.*[111] from Astropower Inc (USA), also reported a similar light trapped, interconnected silicon film. Quin *et al.*[112] reported from Peoples Republic of China, producing a double layer (silicon-oxynitride and silicon-nitride) deposited on silicon solar cell, as antireflection coating. High microwave power and low substrate temperature facilitated the formation of distinct transition regions. Average reflection from such double layer coating at 0.3-0.9 μm was less than 6%, and practical photoconversion efficiency of silicon cells claimed to improve by 45 %. Zhao *et al.*[113] from Photovoltaic Special Research Center, University of New South Wales, Sydney, Australia reported enshrouding the cell surface in thermally grown oxide to reduce detrimental electronic activity of the solar cells and isotropic etching to form hexagonally symmetrical honeycomb surface texture, which boosted cell efficiency to 24.4% for microcrystalline cells and 19.8% for polycrystalline cells. These honeycomb textures reduces reflection loss as well as substantially increases the effective optical thickness of the cell by causing light to be trapped within the cell by total internal reflection.

Interest has also been seen in developing spongy or porous silicon over crystalline silicon solar cell, for better light entrapment as well as surface passivation. Other potential advantages of porous silicon are–antireflection property and light diffuser. However, its major drawbacks are the light absorption wihin the porous layer and both insufficient as well as unstable surface passivation phenomenon. Among various methods available for growing porous silicon, rapid thermal oxidation (RTO), plasma nitridation, and anodic oxidation, have been found[114] potentially interesting pathways with a low thermal budget in common. An improvement in blue response was observed in RTO treatment at high temperature, which is due to creation of an intermediate oxide at the porous Si/Si interface. No passivation effect is observed in the case of nitridation or anodic oxidation. The modified porous material preserves its light diffusing properties and suffers less from light absorption. In other cases, presence of an additional step as well as the fact that refractive index decreases, which is unwanted from the viewpoint of anti-reflection properties of the coating, occur.

At the end, I will advice readers who are interested in more information on growing amorphous and crystalline silicon for solar cell, to consult the review paper of Rech and Wagnee[115] for amorphous silicon and the review article by Bergmann[116], both from Germany, for further information on the subject.

Finally, it is interesting to note that no attempt has been reported so far regarding production of high purity silicon powder by Reicke active metal production technique[821]. Reicke metal powders are highly active powders in the size range 1-2 μm and precipitated in the solution phase by a special reductant-solvent system. The reduction is carried out with an alkali-metal and a solvent whose boiling point exceeds the melting point of the alkali metal. The metal salt to be reduced must also be partially soluble in the solvent and reactions are carried out under argon atmosphere as:

$$MX_n + nK = M^* + n\,KX \quad (M^* = \text{pure active metal powder}).$$

Since these metal powders are active in nature, these can be compacted with little energy. Similarly, $TiCl_4$ can be reduced by $K(BEt_3H)$ in THF to yield ether soluble Ti (stabilized with THF as complex). Other reductant-solvent system includes K and THF, Na and DME (1, 2-dimethoxyethane), Na/K and benzene/toluene. Sometimes the alkali-metal is used in conjunction with an electron-carrier such as naphthalene, or lithium-naphathalenide, or Li-metal. Readers further interested in the subject may consult the book–*Active Metal*–Ed. by Alois Furstner, VCH-Publication (Weinheim, Germany), 1996 ed.

It will be interesting to note here that isotope separation by ligand exchange system (LXS) and ion exchange system (EXS) as has been successfully tried in the separation[123] of ^{65}Cu and ^{63}Cu, has not been reported for separation of the silicon isotopes. By this process, the heavier isotope ^{65}Cu has been enriched at the front part of the cation and anion exchange resin column of ion exchange displacement chromatograph. The heavier isotope ^{65}Cu found preferentially separated into the malate complex in the solution phase. In EXS, where Cu(I)/Cu(II) exchange reaction takes place in the HCl-anion exchange resin, the lighter isotope ^{63}Cu has been observed to enrich at the Cu(I) chloride anion complex side. The separation coefficient for single-step separation in LXS is found to be 2.8×10^{-4} at 40°C, while that of EXS 3.8×10^{-4} at 60°C. These types of isotope separation by ion-exchange chromatography has also been applied to separate isotopes[124] of Li, boron, nitrogen, and even[125] carbon, which occupies the same group in periodic table as silicon. Nevertheless, nuclear size, shape, nuclear mass and nuclear spin towards enrichment factor calculation by theoretical means is an essential exercise before attempt is made for isotopic separation by chemical exchange reaction. The most important equation used for this purpose is Bigeleisen and Mayer equation[126], which states that isotope separation effect in a chemical exchange reaction is proportional to the difference in masses and inversely proportional to the product of masses of the isotopes. Such calculations have not been reported for silicon in literature thus far. Nishizawa *et al.*[127] reported such calculations for zinc isotopic system, and found an unusual separation factor for an odd atomic mass isotop of ^{67}Zn in isotopic separation by liquid-liquid extraction system, using crown ether. Similar is the case with magnesium, strontium, nickel and barium. All the odd atomic mass isotopes in the even atomic number elements have shown different enrichment factor from the values calculated from the equation of $\Delta m/m.m'$ (where Δm = mass difference between the isotopes and $m.m'$ are their individual masses) as calculated by the above Bigeleisen and Mayer method. The observed odd atomic mass effects in this case is large enough to be beyond experimental error. The isotope separation factor of Zn in chemical exchange reaction using cryptand polymer were precisely measured by the above authors, by means of ICP mass spectrometer equipped with 9 collectors as ion detectors. The enrichment factor ($E_{67,66}$) for zinc was found to be $-3.3329(3) \times 10^{-4}$ and that for ^{68}Zn and ^{66}Zn, $1.846(1) \times 10^{-4}$, and that for ^{70}Zn and ^{68}Zn $7.19(2) \times 10^{-4}$. The sum total of vibrational energy shift from one isotope to the other, and their nuclear mass-shift to the enrichment factor of ^{67}Zn was -1.05×10^{-3}, and the contribution of the field shift caused by the nuclear size and shape of the isotope 5.26×10^{-4}. The contribution of the nuclear spin or the hyperfine structure shift to the enrichment factor of ^{67}Zn was small (1.94×10^{-4}).

CHAPTER 5

Silicon Compounds

5.1 INTRODUCTION

Let us first look at the special electronic configuration of silicon atom and the type of bond it forms with other elements, which largely determines stability of the silicon compounds. From size consideration, silicon atom is larger than carbon atom and less electronegative than both carbon and hydrogen. This indicates that polarity of the Si-H bond is opposite to that of C-H bond. Since its atoms have a larger volume, nucleophilic attack on silicon is easier than that on carbon[132] and this susceptibility exists even with Si-H bond as steric hindrance is virtually absent in such a case. This situation further favours formation of a large number of substituted nucleophilic compounds if thermodynamically the bond strength of the new compound is more and no steric hindrance is anticipated in the new compound. Silicon basically forms tetravalent compounds, but hypervalent compounds have also been reported. Coordination number decreases with multiple bond. Because of its unique position in the periodic table (just between metals and nonmetals) silicon forms both inorganic and organic compounds with good stability.

Charge separation in Si-H bond is not as great as other ionic hydrides and thus Si-H bond is considered mostly covalent in nature. Nevertheless its ionic character can be enhanced by incorporating electronegative groups like halogen or electron pulling groups like vinyl, phenyl, or fluoroalkyl on silicon atom, while presence of electron repelling groups like alkyl substituents, decreases ionic nature of the bond. These substituents affects Si-H bond energy and can be seen in shift in vibrational frequency of Si-H bond with the above substituents. Si-C bond which has about 12% ionic nature and Si-Si bond, on the other hand, is much stronger, as evidenced by their formation only at high temperature. Because of this extra energy requirement for formation of these bonds, compounds formed thereby are also expensive to manufacture. Similar is the case with Si-N bond. This bond has exceptional stability and shows inertness (refractive property). These compounds find extensive high temperature applications where no other silicon compounds will be able to exist. There is another class of compounds which along with Si and N atom possess other substituent atoms like hydrogen and carbon, called 'Silatranes'.[133] They exhibit distinct physical, chemical and most important of all compatible physiological properties; accordingly they find many biological applications nowadays.

Silicon-metal bond (e.g. with group I, II and III metals), on the other hand are weaker and can be dissociated easily by polar solvents. There is no known reaction where Si-H bond has been replaced by stable bonds of silicon with elements of the above group metals.

Because of the above basic nature of Si-H and Si-C bond, a large number of carbon family, like organosilicon compounds are formed by these bonds. Similar repeated units of monomers can also be formed with Si-O bond, which are more ionic in nature and consequently have higher bond strength. They show good stability towards heat, UV-radiation, and oxidation resistance. Futher their desired physical properties like flow behaviour, surface modifying ability, hydrophobicity, etc. can be further enhanced by incorporation of suitable substituents into the molecule. Cyclization and bond multiplication of these compounds manifest newer properties like photosensitivity, resinous or rubber-like nature. The silozane bond thus flexes and rotates fairly easily about Si-O axis, especially with small substituents like methyl, on silicon atom.[134] Rotation is also free about the Si-C axis in methyl silicon compounds. As a result of freedom of motion, intermolecular distances between methylsiloxane chains are greater than between hydrocarbons, and intermolecular forces are small.[135] Further these polymers have a unique flexible helical configuration, which allows it to absorb extension due to heat and thus these compounds show little variation in physical properties like viscosity with temperature and good thermal stability. This is in contrast to hydrocarbon polymers which have a stiffer structure.[136]

Silicon, unlike carbon, also possess electrons in *d*-orbitals, which are capable of participating in bond formation with pi-electrons of other foreign atoms which show double bond character.[137-140] Hybridization of this kind gives rise to more covalent character to the resultant bond than ionic, and thus oxygen atom in ethers show more basic nature than the oxygen in silicones. Similarly, chance of electrophilic attack in silicon compounds, e.g. $(CH_3)_3\,SiCH = CH_2$ is half of that on $CH_3\,(CH_2)_3\,CH = CH_2$.[141] In the absence of molecular orbital effect, electronegativity principle explains general reaction tendencies, e.g. greater basic character of nitrogen in $(CH_3)_3\,SiCH_2\,NH_2$ over nitrogen in $CH_3\,(CH_2)_2\,NH_2$.

Silicon atom is also capable of utilizing its vacant 3*d*-orbitals by expanding its valence beyond 4 (to 5 or 6), thereby forming new bonds with electron donors. Existence of hypervalency was reported recently, with isolation of penta-coordinated silicon species with five Si=C bond, by Kolomeitsev *et al.*[142-144] Unlike C=O, the Si=O is unstable, at least at low temperatures. Another unique property of Si is that, with Si-O single bond, it can form a polymeric chain, but carbon is capable of forming only one molecule and not polymeric molecule with oxygen. Silicones thus occupy an intermediate position between organic and inorganic compounds, and in particular between silicates and organic polymers.

One recent development in newer Si-C bond formation has been synthesis of organosilicon derivatives of fullerenes.[145] Fullerenes are football-like structures, being composed of plate-like hexagonal and pentagonal carbon network. Silicon and its derivatives are attached externally to these carbon structures through Si-C bond. Because of this unique configuration, they show various electronic and physico-chemical properties not found with other silicon compounds.

5.2 INORGANIC SILICON COMPOUNDS AND THEIR COMMERCIAL IMPORTANCE

Silicon forms a number of important compounds which are stable and have great commercial importance. Among these, compounds with hydrogen, halogens, sulphur and oxygen are important. Halogen compounds like silicon tetrachloride ($SiCl_4$), dichlorosilane (Si_2Cl_6) or disilane chloride, and intermediate chlorosilanes will be discussed in subsequent chapters in detail, along with their commercial value. Among these chlorosilanes, hexadichlorosilane (Si_2Cl_6) can be produced at a much lower temperature than other chlorosilanes. It is generally prepared by coprecipitation reaction, such as:

$$Si_3Cl_8 + SiCl_4 = 2Si_2Cl_6$$

Its other methods of preparation are from trichlorosilane ($SiHCl_3$) at 0 °C in electric discharge under 100 torr pressure,[146] or by degradation reaction with calcium-silicide. Optical fibers can be prepared by direct surface coating of preforms by flame hydrolysis of Si_2Cl_6.

With oxygen, silicon forms oxides like SiO_2, SiO, and silicon peroxide. SiO is formed as a brown powder when SiO_2 and Si, or SiO_2 and an insufficient amount of carbon for complete reduction are heated above 1250 °C. At lower temperature (600-1000 °C) SiO disproportionates to SiO_2 and Si. It is used in small quantities to produce protective films on semiconductors and as anti-reflective coatings. It is also used on recording tapes. More details on SiO can be found in Chapter 4. Silica (SiO_2) has vast applications, starting as filler compound to water-glass (sodium silicate). I will not describe these conventional items here but interested readers can consult any text book on silica.

Silicon peroxide and its compounds are prepared by reaction of silanes (e.g. tetra alkoxysilanes, tetra aryloxysilane, halosilanes, aminosilanes) with hydrogen peroxide.[147] Gels or powders are produced by this process and later dried. Thus tetra-ethoxysilane (1 mol) was reacted with hydrogen peroxide (2 mol) in aqueous solution with stirring under vacuum and the resulting product was held at atmospheric pressure for 6-12 h, forming a gel containing water and ethanol, which was later dried. Dried product thus produced, contain 35 wt % peroxide groups. Application of these compounds include—as oxidants, bleaching agents, hair bleach, disinfectants, various formulations of cleaning liquid, desulphuriser, drying agent, reducing agent, catalyst, etc.

Another set of silicon-oxygen compounds that are of considerable interest are the hollow spherical aluminosilicate clusters of uniform size and property. These compounds are used as deodorant, catalyst support, humidity controller, etc. They are prepared[148] by the following steps: first simultaneous mixing of solutions containing aluminium compound or transitional metal compounds at a mixing rate of 1 to 10,000 ml/min, or *rapid mixing* of the second solution; then removing the resultant by-product salt, and finally preparing clusters by hydrothermal reaction.

Silicon produces stable compound SiS_2 with sulphur. This compound is being made by thermal decomposition of esters of thiosilicic acid at 300°C:

$$Si\ (SC_2H_5)_4 = SiS_2 + 2\ C_2H_5SC_2H_5$$

The reaction temperature can be lowered to 200 °C by addition of sulphur.

With boron, silicon forms two stable compounds—SiB_3 and SiB_6. Both these compounds are prepared from the basic elements at high temperature in an electric furnace. When an oxygen atom is introduced in between silicon and boron bond, a special class of compounds are formed, which is nick named 'bouncing putty' as they combine opposite properties like, elasticity of a rubber compounds and fluidity of a highly viscous liquid; under slow action of constant force the material runs while under sudden application of force (such as impact with surface) it rebounds like an elastic body. These conpounds are known as borosiloxanes.

With metals of group I and II in the periodic table, silicon forms Si-M bonds and these compounds are known as silicides. Some examples are: Li_2Si, Li_4Si, Li_6Si_2, NaSi, $NaSi_2$, KSi, RbSi and CsSi. They are formed by directly heating the elements at around 600-700°C. These silicides are highly reactive towards water and forms silane instantaneously along with hydrogen.

5.3 ORGANIC COMPOUNDS OF SILICON AND ITS COMMERCIAL IMPORTANCE

Analogous to carbon family, silicon forms a series of organosilicon compounds which because of their bond strength, ability to polymerization, and various desired physico-chemical properties that can be developed by selective substituents, find a large spectrum of commercial usage. In general these compounds are classified into two broad class of compounds- organosilanes (which are derivatives of SiH_4) and silicones (which are siloxanes with repetitive Si-O-Si bond). Organosilicon compounds have at least one organic group attached to silicon atom by Si-C bond. Besides these two main groups of silicon-organic compound, there are also two minor groups, e.g. organometallic compound where metal atom has been incorporated in the polymer chain, and heteroatomic compound where non-metal atom has been incroporated in the silicon-carbon chain.

Following is a brief account of some of the structural units and their special properties as are being used commercially, for preparation of these organic silicon compounds:

(a) Organofunctional silanes

Use: Bonding agent between inorganic materials (e.g. fiberglass, metals, sands, extenders, pigments, etc.) and organic polymers (e.g. polyesters, epoxy resins, synthetic rubber, etc.).

Structure of the compounds

Vinyltrichlorosilane, $CH_2{=}CH{\cdot}SiCl_3$
Vinyltrimethoxysilane, $CH_2{=}CH{\cdot}Si(OCH_3)_3$
Vinyltriethoxysilane, $CH_2{=}CH{\cdot}Si(OC_2H_5)_3$
Vinyl-tris (beta-methoxyethoxy) silane, $CH_2{=}CH{\cdot}Si(OC_2H_4OCH_3)_3$
gama-aminopropyl triethoxysilane, $H_2N(CH_2)_3{\cdot}Si(OC_2H_5)_3$
gama-aminopropyl trimethoxysilane, $H_2N(CH_2)_3{\cdot}Si(OCH)_3$

Imidazolinsilane: ring CH_2—$N(CH_2)_3{\cdot}Si(OC_2H_5)_3$; CH_2 | CH; \N//

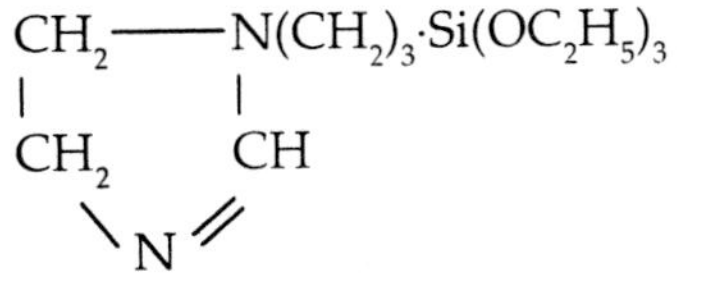

N-minoethylaminopropyltrimethoxysilane, $H_2N(C_2H_5)NH(CH_2)_3{\cdot}Si(OCH_3)_3$
triaminosilane, $H_2N(CH_2)_2NH(CH_2)_2NH(CH_2)_3Si(OCH_3)_3$
gama-chloropropyl-triethoxy silane, $Cl–(CH_2)_3{\cdot}Si(OC_2H_5)_3$
gama-methacryl-oxypropyl-
trimethoxy-silane, $CH_2{=}C(CH_3)COO(CH_2)_3Si(OCH_3)_3$
gama-glycidyl-oxypropyl-
trimethoxy-silane, O (epoxide bridge) CH_2—$CH–CH_2O(CH_2)_3Si(OCH_3)_3$
gama-mercapto-ethyl-trimethoxy-silane, $HS–(CH_2)_3Si(OCH_3)_3$
beta-mercapto-ethyl-triethoxy-silane, $HS–(CH_2)_2Si(OC_2H_5)_3$

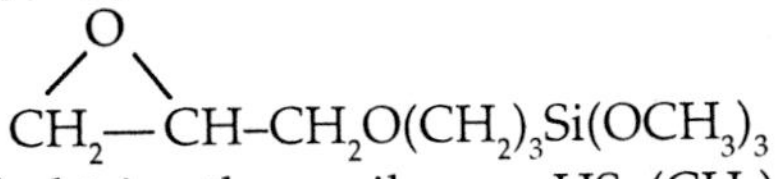

(b) Alkyl silanes

Use: For hydrophobation of mineral fillers and pigments.

Structure of the compounds

n-Propyl-trimethoxy-silane, $CH_3{\cdot}CH_2{\cdot}CH_2{\cdot}Si(OCH_3)_3$
n-Propyl-tris (beta-methoxy-silane), $CH_3{\cdot}CH_2{\cdot}CH_2{\cdot}Si(OC_2H_4OCH_3)_3$
iso-Butyltrimethoxy-silane, $(CH_3)_2CH{\cdot}CH_2{\cdot}Si(OCH_3)_3$

(c) Silicic acid esters

Use: For the manufacture of binders for ceramic materials (precision and investment casting); for surface coating and modifying surface of glasses; for manufacturing anti-corrosion paints.

Structure of the compounds

Tetramethyl orthosilicate, $(CH_3O)_4{\cdot}Si$
Tetraethyl orthosilicate, $(C_2H_5O)_4Si$
Tetra-n-propyl silicate, $(C_3H_7O)_4Si$
Tetramethyl glycosilicate, $(CH_3\text{–}O\text{–}CH_2\text{–}CH_2\text{–}O)_4Si$
Tetraethyl glycosilicate, $(C_2H_5\text{–}O\text{–}CH_2\text{–}CH_2\text{–}O)_4Si$
Silicon aluminium ester, $(C_2H_5O)_3SiOAl(OC_4H_9\text{–}SeK)_2$
Methylpolysilicate, $(CH_3O)_3Si(OSi{\cdot}OCH_3{\cdot}OCH_3)_nOSi(OCH_3)_3$
Ethylpolysilicate, $(C_2H_5O)_3{\cdot}\ Si(OSi{\cdot}\ OC_2H_5{\cdot}\ OC_2H_5)_nOSi(OC_2H_5)_3$

There are also sprayable binders for moulds based on modified silicic esters, especially for zinc rich paints. These binders have higher flash point. Generally ethoxysilanes are used in high temperature zinc rich paints.[150,151] Methyl and phenyl trialkoxysilanes are used for preparation of abrasion resistance coatings on plastics and dielectric coatings for high temperature electronic components.[152,153]

(d) Coupling agents

These are used in molecular level for holding together two dissimilar materials by means of an intermediate compound known as coupling agent. It thus provides interface or bond between matrix and reinforcement in composites.

Chemical compound: These compounds are represented by the general formula $R.Si(CH_2)_nX_3$, where R is an organic radical and X is a readily hydrolysable group. The organic radical may be vinyl, chloropropyl, epoxy, methacrylate, primary amine, diamine, mercapto or a group capable of uniting with a particular resin system. Some examples are–Methacrylate chromic chloride (VOLAN): Coupling agent between polyester-epoxy and phenolic resin.

Silane coupling agent: Used mainly for non-silicious fillers like carbon.

Silanol (a prehydrolyzed silane): Used as coupling agent for glass-fibers.

Trialkoxysilanes (e.g. triethoxysilane and trimethoxysilane): Used for making a number of organic coupling agents utilized by plastic industries as adhesion promoter. Silane coupling agents are also used in solid propellant fueled rocket motors, where it works as a bonding agent, an intermediate layer between insulating layer and the propellant charge[154], thus providing a barrier to migration of nitroglycerine in the propellant. Haloorganosilanes are also useful silane coupling agents.[155]

(e) Metal casting compounds

Ethoxysilanes are used as binders for thin shell making in precision and investment casting. Propoxysilanes are also used in some special applications as silanol formation and set-up time can be easily controlled with this compound. Methoxy-ethoxysilanes are used when a high flash point intermediate is desired, and rates of hydrolysis must remain comparable to ethoxysilanes.

(f) Low heat glasses

These are special glasses being used for manufacturing optical fibers and solar cells, are prepared by cohydrolysing tetramethoxy–, ethoxy–, propoxy–, and butoxysilanes with alkoxides of aluminium, titanium and boron. The process involves slow drying of sols to gels, which are then fired at 500-600 °C to form glass.[156,157]

(g) Hydrophobic agent

Silicon compounds like 2-(Perfluoro alkyl) ethylchlorosilane derivatives serve as both hydrophobic coating and antifouling agent on building surfaces.[158] Water repellant coating for car windshield glass, for masonary work, etc. are formed from methyl–, isobutyl–, amyl–, and octyl–trialkoxysilanes.[159] In masonary work, the mineral surfaces can also be modified with non-functionalized silanes. The main aim in these treatments is to eliminate the polar HOM groups, taking away adsorbed moisture or improve wetting by organic media. Thus hydrophillic surface can be modified by these silicon compounds to hydrophobic and lipophilic. Similarly, thixotropic modification of certain epoxide resins, especially to improve mechanical properties of silicone rubber can be achieved through application of silanized silicas. Condensed product of $Si(OEt)_4$ and $Me_3SiOSiMe_3$ particularly show good result in this type of use.[160] Hydrophobicity to glass apparatus for chemical and medical use as well as in production of light bulb, can be achieved through use of silanization of above glass surfaces. Certain fluoroalkylsilanes have better solvent, dirt, and fat repelling (soil release effect) power than alkylsilyl groups. Volatile siloxanes like $Me_3SiOSiMe_3$, are used to protect vapour deposited aluminium head light reflector and also for microstructure formation on electronic chips.[161]

(h) Special hydraulic fluid

Some of the silicon compounds have very low freezing point (which give them antifreeze property in many applications), as well as good flow property both at high and low temperature. These compounds are successfully used in avionics, high altitude supersonic aircrafts as hydraulic fluid. Example of some of these compounds are Hexakis (2-ethyl butoxy) disiloxane, and Tetrakis (2-ethyl hexoxy) silane. Hydrolytically and thermally stable alkoxysilanes are also used as diffusion pump fluid and provides good service to a vacuum as low as from 1.33×10^{-5} to 1.33×10^{-6} Pa (i.e. from 10^{-7} to 10^{-8} torr) pressure. Example of some of these latter class of compounds are methyltris (tri-sec-butoxysiloxy) silane. Higher tetraalkoxysilanes are used as dielectric fluid, where lubricity requirement and low temperature conditions are present. Airborne radar is one such application.

(i) Analytical applications

Coating of hydrocarbon polymer chain in liquid chromatography is prepared by anchoring octytrialkoxysilanes and other long chain alkyltrialkoxy silanes on substrate material. Many other silicon compounds with similar property are also used for this purpose. These chromatographic columns are used for separating pesticides, protein components, or optical isomers and in ion exchange chromatography. Similar coating of substrate by silylation enables analysis of sensitive natural compounds as well as transient silanols by coupling the chromatographic column, with mass spectrometer (Lenz technique). Even analysis of minerals can be done by such coatings using Me_3Si derivatives.[163] Traces of silicon in the environment can be determined by $Me_3SiOSiMe_3$-catalyst[164] and surface contaminated with silicones can be cleaned using the above mixtures.

(j) Heat transfer fluid

Since many of the silicon compounds have excellent thermal stability, good resistance to oxidation, and high thermal comductivity value, they serve very well as heat transfer fluid in heat exchangers. Example of some such compounds are tetrabutoxysilane, tetrakis (2-ethyl butoxy) silane, and tetrakis (2-ethyl hexoxy) silane.

(k) Application in electronic and photo-optical industry

In micro-electronic application, SiO_2 thin film is deposited on silicon substrate by applying a solution of mixed acyloxysilane to a rotating silicon wafer and then thermally decomposing the silane.[165,166] Fuji-Xerox Corporation, Japan has[167] developed charge transporting agents represented by the general formula $Ar^1Ar^2. N.Ar^5(NAr^3Ar^4)_k. (CH_2)_4SiR^1_{3-a}(OR^2)_a)_1$ where Ar^1-Ar^4 = (un)substituted aryl, Ar^5 = (un)substituted aryl, arylene,

R^1 = H or alkyl or unsubstituted aryl, R^2 = alkyl, n = 2-18, 1 = 1-4, a = 1-3 and k = 0-1. These compounds show good solubility and film forming property and are useful for electroluminescent devices, electroprotog, photoreceptor, etc. prepared by coupling reactions with toluene sulphonic acid esters.

(l) Biological applications

Silanyl-triazines are used as UV-filter in light screen formulations, especially in cosmetics for protecting human skin from sunlight radiation.[168] Some of the silicon derivatives are also used in internal medicines like Cisobitan for the treatment of prostrate cancer, Migugen and Mival as hair growth promoter as well as wound-healing compound. Silatranes show broad-spectrum antibiotic property[169] with specific SiC-bonded groups and opened the door for developing a range of organosilicon pharmaceuticals.[170] Silicon compounds show more specific (site condition in human body) reaction than carbon compounds. In post-ingestion period, hydrolysis of SiOC bonds occur and has been found to degrade more rapidly in the body and thus permits higher dosage with fewer side effects in certain cases.

(m) Silicon compounds as insecticide, herbicide, algaecide, fungicide

A number of silicon compounds have been found as effective insecticide, algaecide, and fungicide. A group of silicon compounds already being commercially sold as fungicide are flusilazole[171] and silafluofen as insecticide.[172] Sagami Chemical Co, Japan also reported a group of siloxanes as bactericides and algaecides, and a group of silacyclohexanes as fungicides (for details of these new compounds, please see Chapter 5, 5.5e subsection). The latter compounds have been claimed to give antifungal property close to commercial flusilazole. Other compounds which have been developed recently having insecticidal, fungicidal and antibacterial property are thio-silicon compounds and vinylsilane compounds of silicon (for more details please see above section 5.5e). Ammonium salt of long chain alkyl substituents to silicon is also used as antimicrobial agent in fabrics such as textiles (carpets, sportswear, etc.), and also for purification of water.[173] These compounds can be represented by the general formula $(RO)_3Si(CH_2)_3N^+Me_3R_1R_2Cl^-$. Fungicidal activity has also been recently reported with S-allyl-o-substituted phenylthiophosphates (–onates).[174]

(n) Synthetic rubber

Polydiorganosiloxanes having large molecular weight, can be converted from highly viscose plastic state to predominantly elastic material by cross-linking (this process is known as vulcanization or curing). These synthetic rubbers are silicon elastomers and have great commercial value for making gaskets for search-lights and turbochargers. The raw material for these viscoelastic materials are gums which are lower molecular weight oligomers with low cross linkages. Silicone rubbers for special specific use (like heat resistance, low temperature, flexibility, etc.) can be achieved by attaching various substituents. Commercially available silicone rubbers have a basic polydimethylsiloxane chain with various types of substituents forming various grades. Generally their molecular weight varies in the range 300,000-700,000. Vulcanization of these rubbers can be achieved both at high temperature or at room temperature, or by chemical means. High temperature cured vulcanizates has better mechanical strength than the room temperature curing. Molecular weight is also a variant towards mechanical performance and best result is obtained with molecular weight[175] around 600,000. Also branching of the basic structure improves low temperature flexibility. Substituents like methyl, naphthyl, benzyl, and phenyl-ether improves low temperature flexibility of these synthetic rubber.[176] Pure polydimethylsiloxane rubber swells in contact with solvents like benzene and petrol, which can be overcome by using fluorosilicone and nitrile silicone rubber.[177] But these substituents also decrease heat resistant property of the polydimethylsiloxane rubber. Instead of vulcanizing these rubbers by mixing with cross-liking agent and catalyst, they can be made into a single pack system as well, which contains cross-linking agent with easily hydrolyzable group. These mixtures cross-link as soon as they come in contact with moisture in the air. One such compound is polydimethylsiloxane-α, ω diols with methyltriacetoxysilane as cross-linking agent[178] or α, ω-bis(methyl diacetoxy siloxy) polydimethyl-siloxane.[179] These rubber compounds also contain structure control additive to impart ultimate strength to the filler network. These are polar type compounds and one such compound is diphenylsilanediol.[180]

(o) Synthetic resin

These are lower molecular weight polymers or oligomers mostly available in liquid form. These are prepared from chlorosilanes and used as impregnating agent, emulsions, adhesion liquid for metal or glass surface, heat resistant fluid, surface coating liquid, lacquers for paints, chemical and solvent resistant layer, water repellant coating, etc. Among all silicone resins, methyl-silicone is the most important one. Because of the low organic content and small space requirement of methyl groups, these resins have hardness on drying and have relatively low thermoplasticity. Incorporation of hydroxyl or alkoxy group bound to the silicon atom, renders the resin either soluble or insoluble in organic solvents. Low temperature curing of resins is accomplished by acyloxy groups.[181] Other curing agent used for these resins are acetyl acetone, acetoacetic ester, malonic acid and biacetyl.[182] Some of the corrosion inhibiting pigments used with silicone resin lacquers are zinc-chromate, basic zinc chromate, serium-chromate, titanium-dioxide, red iron oxide, zinc-oxide and zinc-dust. Other cheaper variety resins that can be blended with silicone-resins are polymethycrylate, UF-resin, melamine formaldehyde, and some polyester and alkyd resins.[183] Foamed silicone resins are used as heat insulating material and as aircraft components. But these materials are brittle in nature and cannot be used as core material for sandwiching elements or areas of vibration. Moulding compounds based on silicone resins contain about 50% filler (inorganic). Laminates made with silicone resins have good thermal stability, low combustibility, and outstanding electrical characteristics even in high humidity; accordingly they find good use in electrical industry. These resins along with glass fibers forms insulating tapes, foils and slot cards. Instead of glass fiber, asbestos fiber is also used sometimes as the matrix element. Electrical properties of these insulators change very little with temperature. Because of their good dielectric breakdown property, they are also used in silicone-isolated machines. Silicone rubbers for their heat resistant property also find some use in electrical industry.

(p) Silicone fluids

Silicone fluids have Si-O-Si units repeated in their polymer and denoted by the general expression MD_xM (where x = 2–4000). Polydimethylsiloxane fluids under this category have superior thermal resistant property (150-250 °C), good low temperature performance (below –70°C), strong hydrophobicity, excellent release property, inert to human body, pronounced surface property, good lubricating property, good dielectric property, little variation of physical properties with temperature change, excellent dampers in vibration, etc. Their viscosity depends on the size of the polymer chain and commercially available fluids have viscosity ranging from 1 to 10^6 mPa. Fluorosilicone fluids show improved low temperature lubricating properties, low solubility in mineral oils and unique properties of fluorosilicone treated surfaces. It is also used as aqueous emulsion for antifoam action, water repellance and release actions.

(q) Silicone greases

These greases have a thick consistency and high temperature resistance property. They are prepared by mixing or dispersing fatty materials of different viscosities to silicone-oil.[184,185] Thus swelling metal soaps in silicone-oil at elevated temperature and subsequently distilling off the solvent results in these speciality grease. Lithium soap (e.g. lithium-stearate and lithium 2-ethyl-hexoate), alkali metal and alkaline-earth metal salts of cycloaliphatically substituted fatty acids and aluminium soaps are used as thickener in these synthetic greases. Other methods followed to produce these greases involve dispersing calcium-stearate in various molecular weight PDMS or polysiloxane, or polymethylphenylsiloxane fluid, or dispersing graphite/carbon-black and copper phthalocyanin, calcium silicate, talc, various organic oxides in colloidal dispersion, or by hydrolyzing organohalosilanes, particularly dimethyldichlorosilane, in presence of highly disperse silica.

(r) Liquid silicone rubber

Preparation of liquid silicon rubber (LSR) has developed relatively recently (early 1980's). These are basically vinyl containing PDMS with viscosity about 1000 times lower than high temperature vulcanizing rubber stock described above. Because of its liquid consistency, it is suitable for processing on injection molding equipment, thus causing lower cost of production of finished article. These products do not contain any solvent and are cured by hydrosilylation reaction. Because it contains vinyl side group, although it does not produce long polymer (in fact, its polymer length is six times lower than HTV-rubber), yet its mechanical and set property is maintained as good as other rubbers, due to extensive cross-linking. Fillers used with this kind of rubber is usually pyrogenic or precipitated silica made hydrophobic by treatment with silylating agents. Fillers give free flowing property to the mass, as these prevent mutual interaction between compounds and thus form localized lumps. These fillers also increase shelf-life of the compound. LSR also has very good curing rate and is capable of producing thin coating on a substrate. It comes in a two-pack system–one containing the polymer with catalyst (usually platinum complex) and the other cross-linker (polymethyl-hydrogen-siloxane) and an optional inhibitor in addition to the base polymer. LSR with other base polymers (e.g. methylphenylsiloxane PVMQ or methyltrifluoropropylsiloxane FVMQ)[186,187] as also with other fillers like carbon black (for electrically conductive tape) are available.[188] Where the material has to be in contact with food-stuff, the cured mass is made further inert (removing last traces of Si-H bond in it) by hydrolysis and partial condensation. General curing of LSR is done at about 170-230 °C (cycle time only 15-60s), while the above refractoriness is developed by heating to 200 °C for about 4 h. Liquid silicone rubber can also be applied on textile fabric to make it waterproof, but the fabric remains permeable to water vapour, a requirement for comfortable feeling with wearable materials.

(s) Silicones for leather industry

Leather is a porous fibrous material which swells in contact with water, but it allows moisture to pass through, thus making wearable items like shoe or garments made from it comfortable. For water repellance treatment of leather, substituted or unsubstituted polymethylsiloxanes, probably almost exclusively are used as silicone treating chemical. Other materials that have been tested but without success are organoalkoxyhalosilanes,[189] polyorganosilazanes,[190,191,192] and organosilanols or polyorganosilaxanols,[193,194] partly because they emanate harmful by-product and partly because they do not produce the desirable effect. Leather impregnated with silicone-oil also passes dynamic testing along with a mixture of polymethylsiloxanes obtained from $(CH_3)_3SiO½$ and $SiO_4/2$ units.[195] A polymethylsiloxane of this type can be made by cohydrolysis of trimethylchlorosilane with $SiCl_4$ or ethyl-silicate in suitable molar ratio. Titanium compounds impart water-tightness to the leather. In this respect, light sensitivity of titanium containing polysiloxane solution can be reduced by addition of aliphatic nitro compounds, e.g. nitroethane or nitrobenzene.[196] Most of the silicone leather impregnating agents available in the market are based on mixtures of polysiloxanes with esters of titanic or zirconic acid.[197] Silatrioxanes and epoxides are being used as tanning agent for leathers,[198] but they are not superior in performance to the conventional tanning compounds (chromium based). A heterocyclic compound having the formula, $Si(CH_3)_2.O.Si(CH_3)_2.CH_2.Sn(CH_3)_2.CH_2$ is also used as antifungal treating agent for leather.[199]

(t) Silicones for paper industry

In paper industry also, silicone compounds are used as water repelling agent and to have non-adhesion effect. At the initial phase of development of these compounds organochlorosilanes[200,201] or alkali metal methylsiliconates[202] were tried successfully for the above effects; but since these compounds liberate acid on hydrolysis and the siliconates were found to have deleterious effect on the fibers, attempts were made to counter these effects by rapid neutralization. Simultaneously less aggressive chemical like polyorganosilazanes[203] and organocycloxysilanes[204] were developed. They are less aggressive, because they produce harmless carboxylic acids. Later attention was turned to a different variety of silicon compounds, siloxanes, which are more readily accessible and easier to handle; one such compound is polymethyl-H-siloxane. Siloxanes with other functional groups also have been found to be suitable for the above purpose (e.g. OH or OR). Still better compounds were later developed, containing a combination of these chemicals along with fluoroalkylsiloxanes.[205-207] But because of cost factor, silicon compounds still cannot compete with its competitor, paraffin waxes, used conventionally for the above effects, although silicon compounds possess some additional advantages like odourlessness, tastelessness, help paper retain its nature, and performs better as release material or covering paper (adhesive foils and tapes). It is also superior in deep freeze packaging material for its superior release action even at a very low temperature. It also acts as a superior release paper in contact with tenaciously adhesive foods, vulcanized rubber, asphalt, pitch, waxes, electrical insulating tapes, adhesive plaster, meat and many other substances. Special compounds are now available for gluing silicon impregnated paper.[208] A thin film of silicon can be transferred to glass by a tissue paper impregnated with liquid polymethylsiloxane.[209] The opposite effect to paper, i.e. to make it adhesive, has also been achieved by use of silicon compounds like polyaminoalkylsilanes,[210] particularly $H_2N.CH_2CH_2.NH(CH_2)_3.Si.(OCH_3)_3$.

(u) Silicones as cleaning and polishing material

Silicones have added advantage as cleaners and polishes for treatment of paints, metal and wood over conventional wax, in the sense that it spreads more easily on the surface and renders the surface hydrophobic and inert. Polishes made by a combination of silicon and wax are also superior in performance over wax alone. For this purpose (like car polish), silicon oil and wax are dissolved in organic solvents or made emulsions (oil-in-water type).[211-215] These silicon oils are generally polydimethylsiloxanes with viscosities in the range of 100-1000 cst.

5.4 DEVELOPMENT OF SPECIFIC PROPERTY THROUGH STRUCTURAL MODIFICATIONS (SUBSTITUENT EFFECT AND FORMULATIONS)

In the last two sections, we have presented various physico-chemical properties silicon compounds possses and commercial value of these compounds. These special properties of silicon compounds arises not only because of Si-C, Si-O or Si-H primary chain in the polymers, but also for the substituents that are attached directly to the silicon atom or to the hydrocarbon side chain. Gross effect of these substituents on the physico-chemical behaviour of silicon compounds have been extensively studied in the past few decades and the knowledge has been used to improvise specific silicon compounds for specific usage. Nevertheless, efficacy of these synthesized, engineered silicon compounds ultimately depends on their actual laboratory tests; however, this information provides only an empirical guideline for engineering silicon compounds. Table 9 below presents some of the effects of the substituent groups on gross nature of the silicon compounds.

Instead of full replacement (e.g. all R) with a particular substituent as shown above, a different or a mixture of groups can be substituted into the same silicon compound, developing new properties. For example, in methylsilicon, if some of the methyl groups are replaced by longer alkyl chains, activation energy for viscous flow and rate of change of viscosity

Table 9: Effect of substituents on properties of silicon compounds.

Substituent	Gross property/application
Alkyl groups	Hydrophobicity. Used as water repelling agent.
Mercapto	Reinforcement property in rubber; polysulphide sealants.
Methyl	Thermal stability; least expensive, release water repellance, gives lowest per cent organic to silicon compound.
Phenyl	Oxidative stability, disrupts crystallinity.
Vinyl	Improves curing; bridge in glass fiber treatment.
Tetrachlorophenyl	Lubricity.
Aminophenyl	Water solubility, bridge to organic materials.
Phenyl-ethyl	Enhances organic compatibility.
Amyl	Enhanced water repellance.
Alkoxy	Gives thermal stability and takes away corrosive chlorine.
Carbethoxyethyl	Bridge to organics.
Fluorine	Imparts high stability to silicones and solvent resistance property.
Amino	Develops thermoset and thermoplastic property; nylon, epoxy, nitrile, phenolic, melamine and other thermosets and thermoplastics.
Diamines	Fiber drawing property. Epoxies, phenolics, melamines, nylon, PVC, urethanes, and acrylic fibers.
Methacryl	Produces polyesters, rubber, polyolefins, acrylics, polysulphide.
Epoxy	Epoxies, urethanes, acrylics, polysulphides.
Modified phenyl	High temperature polymers.
Silazane	Silica treatment for silicon elastomers (wetting property), Novolac, photoresist, adhesion promoter.
All substituent methyl in silicon fluids/elastomers or resins:	Low temperature dielectrics, hydraulic fluid, polishing agent, anti-foaming agent, mould releasing agent, low temperature elastomeric gaskets, coatings and sealants (RTV rubber), pressure sensitive adhesives, prostheses in plastic surgery.
Me/Me + Me/Ph or Ph/Ph in above compounds:	High temperature dielectrics, hydraulic fluid, heat transfer fluid, radiation resistant coating, high temperature flexible or rigid gaskets, insulator coating, HTV rubber sealants.
Me/Me + Me/ $CH_2CH{=}CH_2$ in above Si-compounds:	Coupling agents, cross-linking, copolymerization and reactive centers for further reactions.
Me/Me + Me/ CH_2CH_2X ($X = CN, CF_3$)[2] in above Si-compounds:	Solvent-resistant property, lubrication and sealing property, fire resistant liquids (also by replacing CH_2CH_2X with $C_6H_nX_{5-n}$).

with temperature and pressure increases,[216] while oxidative stability decreases, compatibility with organic compounds become greater and lubricity gets enhanced.[217,218] Instead of using longer alkyl group in the same molecule, if methylphenyl or diphenyl groups are introduced, thermal and oxidative stability increases. Introduction of about 7.5 mole % methylphenylsiloxane into dimethylsiloxane polymer, lowers[219] the pour point from –40 to –112 °C. Here the bulky groups sterically hinders crystallization.

Chlorinated phenyl group (e.g. tetrachlorophenyl siloxane) copolymerized into dimethyl silicon fluid, increases lubricity. The change is a function of the average number of chlorine atoms per phenyl group and the proportion of such groups in the copolymer.[220]

A substituent like a vinyl group, on the other hand, provides reactive centers for further reaction (e.g. cross-linking), or they act as chain propagators when terminally located.

Fluoro compounds like methyl-3,3,3-trifluoro-n-propyl-silicon $(CH_3).(CF_3CH_2CH_2)SiO$, decreases swelling property of silicon elastomers and improves lubricity, but increases rate of change in viscosity with temperature.[221-223] Introduction of polyether substituents also increases viscosity-temperature coefficient, and makes the product more water-soluble. These have good surfactant property and provide the compounds better load-carrying capacity as lubricant.

Incorporation of SiH containing substituent increases its chemical reactivity, which is utilized in textile and paper treatment. It makes the product hydrophobic and its reactivity is utilized in hydrosilyzation curing mechanism. The opposite effect, i.e. drop in reactivity or increase in inertness occurs with substitution of saturated hydrocarbons (e.g. tetraorganosilanes–tetramethyl or tetraethylsilane). These are very stable compounds and have a wide range of melting points.

Reactivity also increases with substituents like unsaturated hydrocarbons (organofunctional silanes). As little as 0.5 % of a hydroxy functional siloxane can increase considerably the ageing resistance of polyurethane elastomers. Reaction of bromo compounds (compared to chloro compounds) proceed more easily.

The dielectric property of silicon oils are considerably modified by nitrile groups, while bulky groups like $(CH_3(C_6H_5)SiO$ inhibits crystallization at low temperature and maintains flow property even at a very low temperature.

In PDMS elastomers, substitution of methyl group with phenyl group improves the low temperature behaviour of the elastomer. However, introduction of phenyl group into the molecule decreases the vulcanization yield as manifested by reduced elasticity, higher tensile strength and higher tear resistance of the vulcanizate. It may also reduce their resistance to swelling in polar and aromatic solvents or naphthalenic solvents and reduce compression set. The phenyl group thus interferes with cross-linking in the elastomers. For a given amount of phenyl group in the polymer, a higher rate of cross-linking is achieved with $CH_3(C_6H_5)SiO$ unit than with $(C_6H_5)_2SiO$ unit.[224] Other substituents like naphthyl, benzyl, and phenylethyl groups may also bring about an improvement in the low temperature flexibility in polymethylsiloxane elastomers containing 2.5 mole % of such substituents. Their effectiveness is in the following decreasing order: alpha-naphthyl, benzyl, beta-phenylethyl, phenyl, alpha-phenylethyl.[225]

Polyethylmethyl siloxanes have vulcanizate with lowest brittle temperature than PDMS elastomers (e.g. up to –136°C). Their rate of crystallization is also lower. The presence of ethyl group in these cases has a similar effect to that of vinyl group (i.e. degree of cross-linking increases, as is the heat stability and elasticity).

Incorporation of heteroatomic groups also improves performance of the silicon rubbers. For example, incorporation of boric acid gives a product which even in pre-vulcanized state, weld firmly to one another.[226,227] The properties of silicon rubber can further be modified by incorporation of organo-silphenylene units into the chain.[228-231]

Introduction of hydrogen atom as substituent to the silicon atom gives vulcanizates with good properties of adherance to metals.[232] Introduction of phenyl group into the organosiloxane network of silicon resin also improves their adhesion property to various substrates. Thus resins with other organic groups, particularly with higher aliphatic groups, have not become technically important, as the phenyl substituted resins. Heat stability also decreases with increase in aliphatic chain length.[240] Alkoxy and hydroxyl groups in such resins (bound to silicon atom) are the reactive centers where condensation reactions leading to the hardening of the resin occurs.[233] Resins with vinyl or allyl[241] groups hardens without catalyst and much faster than resins without alkenyl groups. Presence of such unsaturated groups also increases hardness of the product, as well as their thermal stability.[234,235] Resins with vinyl group also do not emit any volatile matter during hardening and are thus suitable for dipping varnishes.[236,237]

The properties of silicon resins may also be modified by partial replacement of Si-O-Si group by Si-N-Si, Si-Si or Si-C_6H_4-Si bonds. Thus siloxane-silazane copolymers improve adhesion property of the resin to various substrates.[238] Condensate products obtained with organic resins, particularly the products obtained by reaction with polyalcohols, dicarboxylic acids and polyesters containing hydroxyl groups, are not only cheaper but there is also a marked improvement in pigment compatibility, as well as increase in surface hardness and decrease in curing time.[239] Pigment compatibility improves with increasing phenyl to methyl ratio.

Now the extent to which the property of silicones can be changed by substituents alone is a matter of debate, nevertheless attempts have been made to modify properties of silicones by various additives. As for example, pyrogenic or precipitated silica (BET surface area 150-400 m^2/g) has been added to silicon elastomer, which increases its tensile strength as much as 50 times by interacting with the silicon matrix. On the other hand, inert fillers like quartz powder, diatomaceous earth, siliceous and other chalks, talcs, micas, calcium or zirconium silicates and aluminium-trihydrate, decreases hardness and the modulus of silicon elastomer. Active fillers like pyrogenic silica, also increases viscosity due to hydrogen formation in the filler matrix. These newly formed cross-linkages hinder flow even at low concentration in non-polar silicon polymers. Sometimes hydrophobic silylating agents like silylamine and silylacetamide are added to cover filler surfaces and thereby decrease filler inter-particle interactions. Cross-linking can also be inorganic salts of iron, titanium, zirconium, cesium, nickel, copper, cobalt, manganese, and carbon blacks. On the other hand, depolymerization of siloxane polymeric chain can be prevented by use of stabilizers, such as alkaline earth silicates, certain amphoteric hydrides and some organic polymers. Flame retardation to silicones can also be achieved through use of a small quantity (10-60 ppm) of platinum, titanium dioxide, ferrous-oxide and certain nitrogen compounds, which produces a ceramic surface in contact with fire and thereby reduce depolymerization.

Finally, there is another method of modifying properties of silicon compounds by attaching special groups (known as silyl groups) to the polymer itself. These reactions, known as silylation reaction, have assumed ever increasing importance in special, commercially important silicon compounds, and have provided a strong tool to custom tailor silicon molecules for different commercial purposes. The subject being vast, I will discuss these silylation reactions under a separate heading in the following pages.

Silylation reaction

Silylation reaction involves displacement of active hydrogen, like OH, NH, or SH, from an organic molecule by a silyl group. The silylating agent is usually trimethylsilyl-halide or a nitrogen-functional compound. A mixture of silylating agents such as trimethylchlorosilane and hexamethyldisilazone is sometimes used and they are more reactive in combination than either reagent in isolation. Silylation reactions are used in synthesizing product molecules by pharmaceutical companies, where the silylating agents enter the synthesis process temporarily in the production cycle (thus acting temporarily as protector to some useful group in the main molecule). Similarly, in organic synthesis, presence of sensitive protic groups (e.g. OH, NH, COOH, etc.) sometime hinder reaction or render purification more difficult. These problems are circumvented by temporarily substituting the proton by a suitable silylating agent like R_3SiX, where X = halogen, amine or amide. Choice of X groups depend on the possible pH sensitivity of the substrate molecule. Important silane type silylating agents are: Dichlorodimethylsilane (Me_2SiCl_2); Chlorotrimethylsilane (Me_3SiCl); Hexamethyldisilazane $(Me_3Si)_2NH$; N,N′–bis (trimethylsilyl) urea $(Me_3SiNH)_2CO$; N,O–bis (trimethylsilyl) acetamide (CH_3Cl–$NSi(CH_3)_3).OSi(CH_3)_3$; Iodotrimethylsilane Me_3SiI; Hexamethyl-disilane $Me_3Si.SiMe_3$; Tert-butylchlorodimethylsilane $Me_3C.S{\cdot}Me_2Cl$.

The chlorosilanes are generally used in combination with an acid acceptor (e.g triethylamine). After desired reactions have been carried out, the protecting group is carefully removed hy alcoholysis or hydrolysis. To increase the overall economy and decrease environmental pollution, recycling of the free group (e.g. Me_3Si in $Me_3SiOSiMe_3$) is being followed since 1972. Electronic and steric effect of various silyl groups in radical addition reactions have recently been studied by Hwu *et al.*[242] and Hillard *et al.*[243-244]

In pharmaceutical industry, silylation is used in commercial synthesis of penicillin. It is also used to protect a wide range of OH group compounds, encountered in such industries, like alcohols in prostaglandin and steroid synthesis, enols in nucleoside and steroid synthesis, carboxylic as well as sulphonic acids in penicillin and cephalosporin synthesis. Silylating agent like trimethylsilanes are also used extensively in production of completely or semisynthetic betalactone.

Other applications of silylation reaction are, treating minerals like silicates, which are simultaneously end blocked trimethylsilyl under acidic condition to yield specific trimethylsilyl-silicates.[245] These silicates and certain anionic siliconates form stable solutions in water and alcohols, at any pH, and used as corrosion inhibitor in glycol antifreeze.[246] Allyl-silylating agents also make mineral surfaces water repellant, which are useful in masonary work, and electrical work, packings for chromatography and in non-caking fire extinguisher development. A series of organofunctional silylating agents is also used commercially as adhesion promoter (e.g. coupling agent between organic and inorganic particles). Organofunctional silanes are also used to control orientation of liquid crystals (for more detail see Section 5.5), bind heavy metal ions, immobilize enzyme and cell organelles, modify metal oxide electrodes, surface binding of microbial agents, and other non-plastic applications. To achieve a clear display in liquid crystal panel, crystals are oriented perpendicular to the substrate by treating oxide surface with octadecyl-3 (trimethoxysilyl) propyl ammonium chloride, $C_{18}H_{37}.\dot{N}.(CH_3)_2(CH_2)_3Si(OCH_3)_3Cl^-$, or oriented parallel by treatment with N-methylaminopropyltrimethoxysilane[247]. $CH_3NH(CH_2)_3Si(OH_3)_3$.

The ethylenediamine(en)-functional silane has been studied extensively as silylating agent on silica-gel to preconcentrate polyvalent anions and cations from dilute aqueous solutions.[248-250] Electron-transfer reactions can be carried out on underlying metal oxide layer in special electrodes, coated with a monolayer of gamma (beta-aminoethyl) aminopropyltrimethoxy-silane by dipping in benzene solution of the silane.[251] Dye molecules attached to silylated electrodes absorb light coincident with the absorption spectrum of the dye, which is the first step towards simple production of photo-electrochemical devices.[252]

Surface bonded organosilicon quarternary ammonium chloride have been used to enhance antimicrobial and algicidal activity on various surfaces (e.g. 3-(trimethoxysilyl) propyldimethyloctadecylammonium chloride. This property of the chemical has been attributed to a surface bonded chemical.[253] Similarly, (chloromethyl) phenylethylsilane and its derivatives have been used in polypeptide synthesis.[254,255] Silylation has also been helpful in enzyme catalyzed reactions in cell-free systems.[256]

Besides the above, a large number of silylation reactions have been pursued in recent times, and readers interested in these reactions may see Refs (257-366) given at the end of this book.

5.5 CURRENT RESEARCH FOCUS/TREND

Current research trend has been of synthesizing silicon compounds that are mostly of commercial value. Many of these methods of synthesis have been challenging. However, many complicated molecules were successfully synthesized by easy and innovative route. Some fundamental studies (including theoretical quantum chemical studies) has also been performed for some of the intricate molecules. Compounds of commercial value includes industrial materials, insecticide, bactericide, fungicide, compounds for medical and analytical applications, and photo-optical as well as electronic materials. New synthesis route includes electrosynthesis, molten salt route, and various catalytic routes. Design of synthetic experiments have also been improvised through new emerging fields like combinatorial chemistry and high speed computers. We will discusss these subjects in the following pages, but first let us look at the new innovations that have been claimed in recent times.

(a) New discoveries

Successful synthesis of new silicon-compounds and processes that have been reported in recent times (especially in the last three years) are as follows:

- Formation of new cyclotrisilanes and siliconium ion through reaction of cyclotrisilane with Lewis acid, reported by Belzner *et al.*[367]
- Synthesis of new functional siloxane, (8-methoxy-napthyl) silyltriflates, claimed by Castel *et al.*[368]

- New technique for introducing silyl group into alpha, beta-enones using a disilane catalyzed reaction with copper salt was reported by Ito *et al.*[369]
- A new method for generation of dimethyl-silanone under mild condition was reported from Russia by Voronkov *et al.*[370]
- Synthesis of the first stable cyclo-trisilene was done by Iwamoto *et al.*[371]
- Synthesis of the first Si-H-B bridge by combining 1, 1-organoboration and hydrosilylation technique was reported by Wrackmeyer *et al.*[372]
- Synthesis and characterization of first allenic compound with double bonded phosphorus and silicon, Ar.P:C:Si(Ph)TiP was reported by Renaivonjatova *et al.*[373] The S_Hi reaction at silicon: a new entry into cyclic alkoxysilanes was reported by Studer and Stean.[374]
- The first detection of molecules (HCSiF and HCSiCl) with C:Si triple bond, were reported by Karni *et al.*[375]
- New synthetic route for silylcyclopropanols via titanium mediated coupling of vinylsilanes and esters, was reported by Mizojiri *et al.*[376]
- New di-cationic silicon complexes with N-methylimidazole were reported by Hensen *et al.*[377]
- A new method for synthesizing hexamethyl-disilazane was reported from China by Xu *et al.*[378]
- High pressure synthesis of new silicon containing heteroatom analog of fused norbornenes was reported by Kirin *et al.*[379]
- A new class of potent protease inhibitors (silanedioles) was disclosed by Sicburh *et al.*[380]
- First exclusive endo-dig carbocyclization: $HfCl_4$ catalyzed intramolecular allylsilation of alkynes was reported by Imamura *et al.*[381]
- Synthesis of new cyclolinear permethyloligosilane-siloxane was reported from Russia by Chernyavaskaya *et al.*[382]
- Synthesis of a first kinetically stable dibenzosilafulvene was also from Russia by Zemlynski *et al.*[383]
- A group of iminosilanes which can be used as precursors for new rings and also for unknown ring systems are reported by Klingebiel and Niermann.[384]
- A new class of cyclic stannyl-oligosilane is reported by Kayser *et al.*[385]
- Synthesis and catalysts for new organosilyl-transition metal complexes (the activation of Si-Si sigma bond by transition metal complexes) is reported by Suginome and Ito.[386]
- A group of new dichlorosilanes, cyclotrisilanes, and silacyclopropane as precursors of intermolecularly coordinated silylenes is reported from Europe by Belzner *et al.*[387]
- First experimental evidence was put forward by Clannes and Dillon,[388] regarding formation of a silicate anion by intramolecular addition of a persulphoxide to a trimethylsiloxy group.

- Synthesis of the first phthalocyanine containing dendrimer was reported by Kraus and Lousw.[389]
- A new class of chelating silylamido ligands for synthesis of lithium and magnesium derivative of (t-BuHSiMe_2–O–C_6H_4X) where, X = OmE, NMe_2, CH_2NMe_2 and CF_3 was reported by Goldfuss *et al.*[390]
- Present new trends of acylsilane chemistry was described from Italy by Bonini *et al.*[391]
- A new and easy route of synthesis for bis (trimethylsilyl) ketone was reported by Pan and Bennecte.[392]
- A new method for following the kinetics of the hydrolysis and condensation of silanes was described by Lindburg *et al.*[393]
- Formation as well as properties of a new class of sila-heterocycles were presented by Kroke and Weidenbruch[394] at Organosilicon Chemistry, 3rd International Conference in 1996.
- A new route to silaheterocycles: nucleophilic aminomethylation was also presented at the above International Conference by Karsch and Schreiber.[395]
- Controlled cleavage of $R_8Si_8O_{12}$ framework: a revolutionary new method for manufacturing precursors to hybrid inorganic-organic material was described by Fehr *et al.*[396]
- A new and easy route for manufacturing polysilanyl-potassium compound was reported by Marschner.[397]
- Functionalized trisilylmethanes and trisilylsilanes as precursors for a new class of tripodal amido ligands were reported by Schubart *et al.*[398]
- Tararov *et al.*[399] described the first structurally defined catalyst for the asymmetric addition of trimethylsilylcyanide to benzaldehyde.
- A process was described by Jeschke *et al.*[400] for the synthesis of first known germinal-di(hypersilyl) compound, Methoxy(bis)(tris) (trimethylsilyl) methane.
- A new example of linear disiloxane: synthesis and X-ray crystal structure of bis (2-silolyl)tetramethyl-disiloxane was reported by Yamaguchi *et al.*[401]
- A novel organo-lead(II) compound with silicon, its synthesis and structure, was described by Eaborn *et al.*[402]
- A new class of magnetic organometallosiloxane was reported from Russia by Leviskii and Buchaecenko.[403]
- A new class of water soluble organosilane compounds which can be used as a radical reducing agent in aquous media was reported from Japan by Yamazaki *et al.*[404]
- Synthesis of a new spirocyclic system, silaspirotropylidene was published by Sohn *et al.*[405]
- For the first time synthesis and characterization of the stable heteroleptic silylstannylenes was reported by Drost *et al.*[406]

- Method of preparation for a new class of silane compounds which can be used as coupling agent with inorganic particles was reported from Japan by Kawai and Nekanchi.[407]
- Bonini *et al.* has published a very interesting account of the new chemistry of alpha-silylvinyl-sulphides.[408]
- Possibilities of synthesizing precursors for new materials from silicate cage compounds were described by Harrison.[409]
- Process for synthesizing new functionalized bis-acylsilanes was reported by Boullion and Portella.[410]
- Silacyclopentadienylidene, the first silylene incorporated in a silol ring, its synthesis and structure was described by Kako *et al.*[411]
- Narula *et al.*[412] from India, described the first silatrene with a direct Si-NCS bond (the 1-ioscyanato-Silatrene).
- A new synthetic route to allylsilanes was described by Saito *et al.*[413]
- Due to limitation in space and scope of the book, the above findings and disclosures are not described in detail; but interested readers can consult the relevant references provided for further details.

(b) Fullerene–Silicon Compounds

With its discovery in late eighties and subsequent breakthrough in producing a reasonable quantity of fullerene for synthesis work, attempts were made in the last few years for synthesis of new compounds, hitherto unknown, by reacting fullerene and silicon compounds. Interest in these compounds can be traced to the fact that buckyball-shaped carbon derivative, C_{60} fullerene, offers an array of labile pi-electrons in a three-dimensional space, which can form *d-p* double-bond like hybrid linkage with silicon atom. While in Inorganic Chemistry, interest lies in incorporating the metal atoms into the inner space of the buckyball C_{60} fullerene, for organic chemists interest was to anchor the organosilicon compound outside the cage through carbon-silicon *d-p* bond. These efforts have paid through discoveries of new silicon compounds which can be used as wavelength converter, optical switches, etc. Although existence of higher fullerenes (e.g. C_{120}, etc.) are also known, from stability and ease of production considerations, investigations of compounds formed with C_{60} variety of fullerene only are of commercial interest. We will discuss some of these recently reported compounds, to elucidate the type of work continuing in this area and the future possibilities.

At the University of Tsukuba, Ibaraki (Japan), photochemical-functionalization of C_{60} fullerene with phenylpolysilane has been studied by Kusukawa and Ando.[414] In particular, they studied photolysis of tertiary-butyl substituted disilane with C_{60}-fullerene, which resulted in the synthesis of the following 1, 16-adduct; where the silyl and phenolic groups attached to 1, 2-position of fullerenes are also obtained from the reaction of

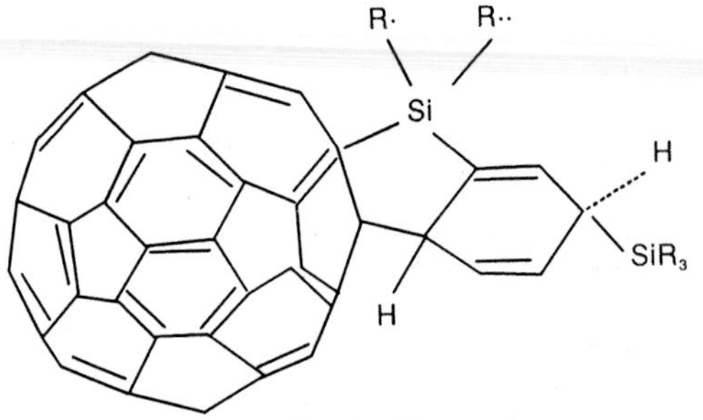

Product I

silanes, such as $(TMS)_2$-Ph.Si.Si.Ph$(TMS)_2$ and $(R_3Si)_3(2,3,4\text{-}R'_3\text{–}C_6H_2)$, where R = methyl or ethyl group and R´ = H or methyl group. Structure of all these compounds were determined by the above authors and represented through one-and two-dimensional NMR.

At Nippon Kayaku Co, Japan, Yoshida and Mori[415] prepared fullerenes with silyl-groups having improved optical functions. Compounds of general formula–H_nC_m $[(SiRR')_pR'']_{n'}$, where C_m = residue of fullerene with C number m, whose n double bonds have been converted into single bonds; $m \geq 1$; n,q (average number) ≥ 1; R–R´´ = C_{1-18} (un)substituted alkyl, C_{6-18} (un)substituted aryl, C_{1-8} alkoxy, OH, and halol were prepared, which are useful as wavelength converters, optical switches, etc. In these experiments, 1.44 mg of C_{60}-fullerene was treated with 2.12 g of Ph_2SiHCl and chloroplatinic acid in MePh at 70 °C for 3 h to produce 1.65 g of Ph_2SiCl modified C_{60}-fullerene.

In Russia, Bespolova *et al.*[416] studied hydrosilylation products of C_{60}-fullerene. In these experiments, hydrosilylation of C_{60}-fullerene by diphenylsilane was studied, whereby silicon containing derivatives like $C_{60}H_n(HSiPh_2)_n$ (where, n = 2,4,6) were obtained. These products are most probably 1,2-addition derivatives of Ph_2SiH_2 at active (6,6)-double bonds of C_{60}-fullerene. The formation of 1,4-isomers require higher energy, because the localization of double bonds in the fullerene-C_{60} ring changes in such a way that the (6,5) edge becomes doubly bonded. An excellent review article on novel fullerene based organosilicon compounds has been written by Akasaka *et al.*,[417] summarizing the photochemical bis-silylation of fullerenes (citing 20 references) has been published recently, and interested readers on this subject can refer to the above reference for further details on the subject.

From Madrid, Spain, Iglesias and Santos[418] reported reaction of fullerenols, $C_{60}(OH)_x$, where x = 12 or 8, with trialkoxysilane, $(RO)_3Si(CH_2)_3X$ where R = methyl and X = Cl, and when R = ethyl, X = NH_2. In these transesterification reactions of trialkoxysilanes, the $SiR(OR')_3$ where R = $(CH_2)_3Cl$ and R´ = CH_3, and when R = $(CH_2)_3NH_2$, R´ = ethyl, with fullerenols $C_{60}(OH)_x$, produced various products containing totally or partially hydrolyzed $OSiR(OR')_2$, $O_2SiR(OR')$ and possibly O_3SiR attached species in a variable number. It was reported that the yields with respect to the fullerenols in such reactions were very high (33-99 %).

At Nigata University, Ikarashi (Japan), Akasaka *et al.*[419] studied the chemical derivatization of fullerene with organosilicon compounds. Their studies involved irradiation of benzene solution of an equimolar mixture of C_{60}-fullerene with disilane (tert-$BuPh_2$.Si)$_2$ in a quartz tube with a low pressure mercury are lamp, which resulted in product 1,6-adduct of C_{60}(tert-$BuPh_2Si)_2$. The product seems to be easily separable by preparative HPLC, and its analysis indicates a structure with C_2-symmetry. The product is also found to have a significantly lower oxidation potential compared to some analogous cyclic 1, 2 and 1, 4 adduct. Of the nine possible isomers with C_2-symmetry possible with related type $C_{60}(SiH_3)_2$ compound, the 1, 2-adduct was found to be the most stable. An excellent review paper on a related subject, '*Organosilicon derivative of fullerene*' has recently been published by Ando and Kasakawa[420] from University of Tsukuba, Japan (containing 30 related references). Readers interested in further details on the subject may consult the above review paper.

At CNRS, France, Ray *et al.*[421] reported synthesis and determination of structure of silicon-doped hetero-fullerenes. They synthesized clusters of $C_{2n-q}Si_q$ (where $2n = 32–100$ and q less than 4) by this method. Analysis of the abundance distribution and photo-fragmentation spectra provided clear evidence that such clusters remain in the fullerene geometry, and that silicon atoms are located close to each other in the fullerene network. In above paper, authors also discussed stability and electronic properties of these heterofullerenes.

Apart from the above study, substituent effect on addition of silyl-lithium to C_{60}-fullerene and also germyl-lithium to C_{60}-fullerene, have been studied by Kusukawa and Ando,[422] and exohedral modification of fullerene-C_{60} with several silicon containing adducts (together with synthesis procedure and their characterization) has been reported by Miller.[423]

Some of these silicon-fullerene compounds have shown extraordinary properties with respect to both electronic effect and mechanical effect, and some of them have potential for commercialization in critical applications in the near future.

(c) Liquid crystals

Liquid crystals and its precursors have been synthesized in recent years from organometallic silicon compounds. These compounds are suitable for making homotropically oriented liquid crystalline phases on surfaces such as glass plates, and coatings on them. These materials are useful for preparation of electro-optical indicating elements, due to the negative dielectric anisotropy of such oriented liquid crystal phases in the applied external electric field. One such example of these class of compounds is represented by the general formula (reported by Panluth *et al.*[424]) $R(A_1)_m Z_1 A_2 (Z_2 A_3)_n WSiX_a Y_b Z_c$ (abbreviations of the letters described below). One such particular compound, *p*-$PrC_6H_{10}C_6H_4.(CH_2)_4Si(OMe)_2$ where C_6H_{10} represents 1,4 cyclohexylene, was coated on ITO glass in 6,12, and 24 µm thickness with DVK K-30 binder in ethanol. The coated places on the ITO-glass was found to show the desired orientation of the molecular crystals. Several such trimethoxysilyl compound were prepared by treating the corresponding terminal alkenyl compounds with $(MeO)_3SiH$ in presence of chloroplatinic acid (H_2PtCl_6) in the mixed solvent of CH_2Cl_2–Me_2CHOH. The general representation of these compounds as shown above have the following abbreviations: R indicates (a) chiral C_{1-10} alkyl in which one or two non-neighbouring CH_2 can be replaced by oxygen, fluorine or chlorine atom, halogenated C_{1-3} alkyl, alkoxy, alkenyl or alkenyloxy group. A_1, A_2, A_3 represents 1,4-cyclohexylene, in which one or two CH_2 can be replaced by oxygen and/or sulphur atom, 1,4-cyclohexylene, 1,4-phenylene, in which one or more CH can be replaced by N; any of the above A_1, A_2, A_3 can also be substituted by one or two fluorine atoms. Z_1 and Z_2 represents CH_2CH_2, CO_2, C_2H_4, $(CH_2)_4CH_2CHCHCH_2$, CH_2O, OCH_2, CH:CH or C⋮C (triple bond). The integer m in the above expression is either 0 or 1, and $n = 0$–2, but $m + n \geq 1$. W represents (un)branched C_{1-10} alkylene, in which one or two non-neighbouring CH_2 can be replaced by CHF and/or $CHCF_3$. X,Y,Z indicates OCN, CN, R, OR, H, Cl with at least one X, Y, or Z (H). R indicates C_{1-15} alkyl, in which one or more non-neighbouring CH_2 can be replaced by O, CO, and/or CH:CH. a,b,c, = 1,2,3 with $a+b+c = 3$.

Takeda *et al.*[425] reported preparation of silacyclo-hexanones and their intermediate silanes, as intermediate for liquid-crystal making. In this particular study, the silacyclohexanone (shown in the following figure) where Ar = H, Ph, tolyl, C_{1-10} linear alkyl, C_{3-8} branched alkyl, C_{2-7} alkoxyalkyl,

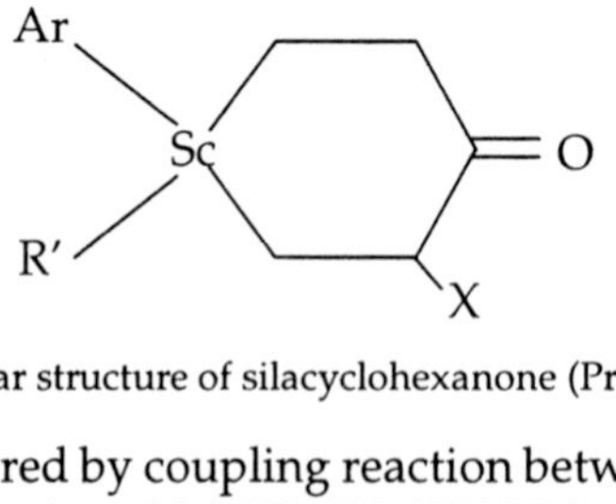

Molecular structure of silacyclohexanone (Product–I)

and X = H, were prepared by coupling reaction between $Ar.R'.SiCl_2$ (Ar, R′ being the same as above) and $MCH_2CH_2CH(OR_2)_2$.

The other intermediates produced in the process are:

Product II: M = Li, MgX, ZnX; X = halogen; $R^2 = C_{1-5}$ linear alkyl, C_{3-6} branched alkyl; $R^2R^2 = C_{2-3}$ alkylene, by oxidation of $ArR'Si[CH_2CH_2CH(OR^2)_2]_2$.

Product III: Ar, R′, R^2 = same as above, by H_2O_2 and hydrogen-halides in alcohols, intramolecular cyclocondensation of Ar.R′.Si $(CH_2CH_2CO_2R^3)_2$.

Product IV: Ar,R′ = same as product I; R^3 = C_{1-5} linear alkyl, C_{3-6} branched alkyl, prepared with bases and dealkoxycarbonylation of product I; Here X = $COOR^3$.

Thus, $PhPrSiCl_2$ was reacted with product II (having M = MgBr and R^2R^2 = $(CH_2)_2$ in THF at room temperature for one hour to yield 96 % of product III (with Ar = Ph, R′ = Pr, R^2R^2 = $(CH_2)_2$), which in-turn was oxidized by hydrogen peroxide and hydrochloric-acid in methanol to produce 91 % of the product IV (having Ar = Ph, R′ = Pr and R^3 = CH_3). Reaction of the ester with NaH in Ph.Me at 100 °C for 3 h and subsequent heating with NaCl under DMSO reflux produced 91 % of the product I (having Ar = Ph, R′ = Pr and X = H), which was finally converted to the liquid-crystal 4-(trans-4-propyl-silacyclohexyl)-4-fluorobiphenyl.

The same group of scientists at Shin-Etsu Chemical Co, Japan, in an earlier communication[426] did report a similar group of silicon compounds for liquid-crystal making. In this particular study, the product silacyclohexanone as shown in the above figure (where Ar = Ph, tolyl; R′ = Ar, C_{1-10} linear alkyl, C_{3-8} branched alkyl, C_{1-10} mono or di-fluoroalkyl, C_{2-7} alkoxyalkyl, and X = H) were prepared by the following four methods:

(i) Reaction of $Ar.R.SiH_2$ (Ar, R = same as above) with CH_2 : CHZ (where Z = CO_2R', cyano and R′= C_{1-4} linear alkyl, C_{3-5} branched alkyl) in the presence of transition metal catalysts.

(ii) Hydrogenating mixtures of Ar.R.Si(CH=CHZ), $ArRSi(CH_2CH_2Z)$ CH:CHZ, and ArRSi, $(CH_2CH_2Z)_2$ (where Ar, R, Z = same as above).

(iii) Converting $ArRSi(CH_2CH_2Z)_2$ into compound (Product I) above, where X = CO_2R' and R′ = same as above, by either of the following:
- treating with bases (when Z = CO_2R^2), or
- hydrolyzing with acids, esterification, and condensation (when Z = cyano), or
- cyclocondensation by bases and hydrolysis with acids (when Z = cyano).

(iv) Dealkoxycarbonylation or decarboxylation: $Ph.Pr.SiH_2$ (7.5 g) was reacted with 25 g ethylacrylate in benzene using $Co_2(CO)_8$ at room temperature for one hour, and later hydrogenated in benzene using $(P.Ph_3)_3RhCl$ at room temperature to produce 3.8 g $Ph.Pr.Si(CH_2CH_2COOC_2H_5)_2$.

As above, treating 19.4 g of the diester with NaH in Ph.Me at 100°C for three hours and refluxing in DMSO-water in presence of NaCl, for 15 h, produced 11.5 g of Product I (with Ar = Ph, R = Pr, X = H). This product was later converted into the liquid-crystal 4′-(trans-4-propyl-4-silacyclohexyl)-4-fluoro-biphenyl.

Another patent filed by the same company (Shin-Etsu Chemical Industries, Japan) almost at the same period, in the same line, where the silacyclohexanone with molecular structure as Product I, shown above (with Ar = Ph, tolyl; R´ = Ph, tolyl, C_{1-10}(perfluoronited) alkyl, C_{2-7} alkoxyalkyl; R_1 = H) were inexpensively and safely prepared with high yield, by treating $ArSiRX_2$ (with Ar and R same as above and, X = Cl, Br, trifluoromethanesulphonate group, alkoxy) with HC⁝CM (M = Li, Na, MgY; Y = Cl, Br, I), and treating the product ArSiR, $(C⁝CH)_2$ (where Ar and R are same as above) with bases and carbon-dioxide gas, followed by hydrogenation, esterification, cyclization, and dealkoxycarbonylation. Thus, reaction of 60 g $CH_3(CH_2)_4SiPh(CH_2CH_2COOC_2H_5)_2$ with NaH in MePh at 100°C for three hours yielded 58.6 g crude product I (where Ar = Ph, R = $CH_3(CH_2)_4$ and R´ = $COOC_2H_5$) which was refluxed with NaCl in DMSO-water for five hours to produce 36.3 g of product I (with Ar = Ph, R = $CH_3(CH_2)_4$, and R´ = H). The liquid crystal 4´-(trans-4-nPentyl-4-Silacyclohexyl)-4-fluorobiphenyl was then formed via Grignard reaction of compound I with 4-bromo-4´-fluorobiphenyl.

Besides the above compounds of commercial interest, synthesis of (fluoroalkoxy) alkoxysilane as material for SiOF interlayer insulator film, for semiconductor devices, prepared by CVD technique has been described by Hijido *et al.*[428]

(d) Photofunctional silicon-compounds and photo-induced reactions

A good number of publications have appeared in literature in recent time, involving studies on photophysics and photochemistry of silicon compounds. A short account of these studies is described below.

At Korean Advance Institute of Science and Technology, Shim and Park[429] studied photo-induced intramolecular cyclization of 1-(0-allyloxyphenyl)-2-pentamethyldisilanylethyne having molecular formula, 2-$R_2C{:}CHCH_2OC_6H_4C⁝CSiMe_2$. $SiMe_3$ (where, R = H or CH_3) in benzene solution, generated the novel stereo-selective intramolecular cyclization product with following molecular structure:

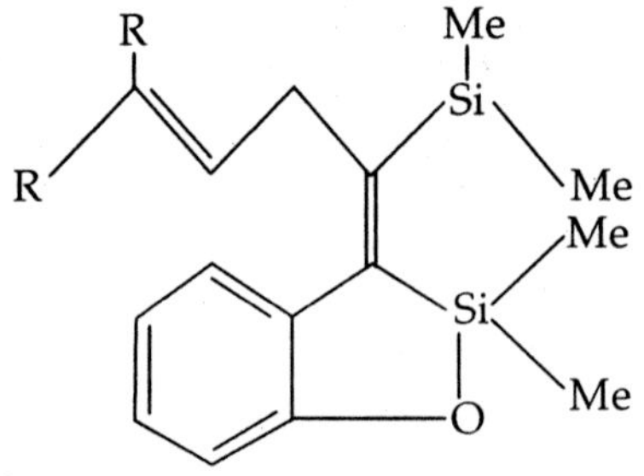

The same workers reported,[430] photoreaction of 1-(0-acetoxyphenyl)-2-pentamethyl-disilanylethyne. Here, irradiation of 1-(0-acetoxyphenyl)-2-

pentamethyldisilanyl-ethyne and 2-$CH_3CO_2C_6H_4C{:}CSi(CH_3)_2Si(CH_3)_3$, in benzene yielded the *photo-Fries rearrangement products* 1-HO-2-Ac-6-$C_6H_3C{:}CSiMe_2R$ where R = Me or $SiMe_3$, and a photo product $2MeCO_2C_6H_4C{:}CSiMe_3$ via silacyclo intermediate.

Preparation of tetrathienylsilane derivatives which are used in electrical, optical or magnetic devices has been described by Nakayuma *et al.*[431] The parent compound, whose derivatives are used for such electrophotogenerating, photoreception, in organic electroluminiscent devices, nonlinear optical devices, etc., by themselves or as their polymer, have the following general structure; where R^1, R^2, R^3 and R^4 are alkyls, (un)substituted aryls or (un)substituted thienyl.

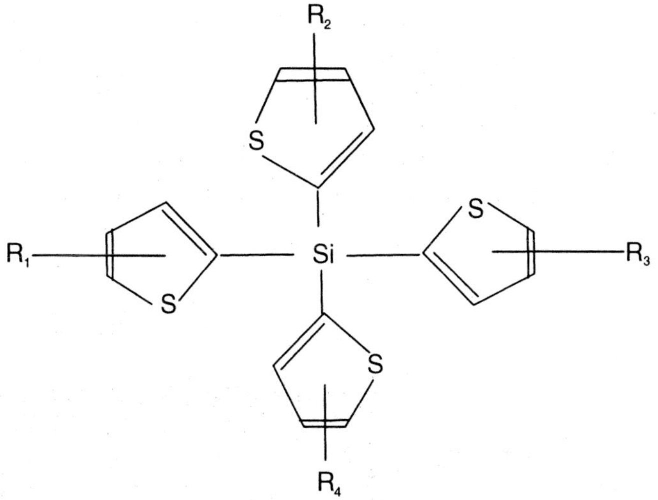

These compounds are prepared by treating Si $(OMe)_4$ or trithienylmethoxysilane-derivatives with thienyl-lithium derivatives in organic solvent at a lower temperature.

Nugent[432] at E.I. DuPont DeNemours (USA) disclosed a process for manufacturing optically active halohydrintrialkylsilylether. Here, the active compound beta-halohydrin (protected as their trimethylsilylether derivative, whose molecular structure is shown below), was prepared using enantio-selective Lewis-acid hafnium (Hf) or zirconium (Zr) catalyst complexes. In the above structure, R = substituted hydrocarbyl or the two R´s form a ring, and X = Cl, Br, or I. The complex comprises optically tri-isopropanolamine of (R,R,R)-tri-isopropanol-amine with Zr(IV) tertbutoxide in THF, followed by treatment with water and trimethylsilyltrifluoroacetate which produced (R)-catalyst. (R)-catalyst catalyzed reaction of aziodtrimethylsilane with 1, 5-cyclo-octadiene monoepoxide and allyl-

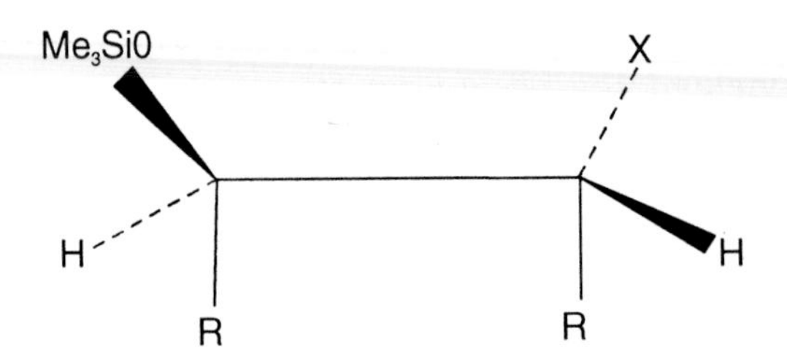

bromide in the presence of chlorobenzene produced 72 % (1R, 2R)-1-bromo-2-trimethyl-siloxycyclooct-5-ene in 79 % enantiometric excess.

Yamada *et al.*[433] at Fuji-Xerox Corporation, Japan, reported preparation of photofunctional hydrolyzable organosilicon compounds. These compounds, represented by the general formula–$A[CH{:}CHYSiR^1_{3-n}(OR^2)_a]_n$ where R^1 = H, alkyl, arlyl; R^2 = H, alkyl, trialkylsilyl; A = a photofunctional group; Y= divalent group; a = 1-3; and n = 1-4, is useful in making organic electronic devices (e.g. electrophogenerating devices). These compounds are prepared by reaction between $M(R^3)_2P^+.hal^-CH_2YSiR^1_{3-a}(OR^2)$, where R^3 = alkyl or Ph; hal = halogen group atom, M = support, and Y, R^1, R^2 as well as a represent the same as mentioned above, with $A(CHO)_n$ (A and n also are same as above). Polymer supported PPh_3 (25 g) was quarternized by 25 g of $(CH_2)_3Si(OMe)_3$ to produce 40 g phosphonium salt supported on polymer. 5 g of this later compound was then treated with NaH followed by reaction with 5 g of N-(4-formylphenyl)-N-(3,4-dimethylphenyl) biphenyl-4-amine in DMF at 70°C for five hours to yield 5.1 g of the desired product (title compound with A = C_6H_4-p-N($C_{12}H_9$)$C_6H_3Me_2$-3,4; $C_{12}H_9$ = 4-biphenyl, Y = $(CH_2)_2$, R^2 = Me, a = 3, and n = 1).

From Russia, Bolshovkova *et al.*[434] reported sulphochlorination of tetramethylsilane and hexamethyldisiloxane. In particular, reaction of the sulphurylchloride with tetramethylsilane and hexamethyldisiloxane, in the presence of photocatalyst Y(III) chloride and sulphur, through hydrolysis and ammonolysis of chlorosulphonylmethyl (trimethyl) silane, produced trimethylsilylmethane sulphonic acid and its amide.

Kako *et al.*[435] from Japan, reported results of their studies on photoinduced electron transfer reaction of 7,8-disilabicyclo (2,2,2) octa-2,5-diene. In this case, 9,10-dicyanoanthracene-sensitized irradiation of 7,8-disilabicyclo (2,2,2) octa-2,5-dienes in the presence of methanol, resulted in the formation of dimethoxydisilanes and the corresponding aromatic-compounds, 9,10-dicyanoanthracene sensitized photoinduced reaction of disilabicyclooctadiene in methanol/CH_2Cl_2 produced MeOSi(i-Pr_2) Si(i-Pr_2)OMe, anthracene and silylanthracene dimer. Crystal structure of the dimer was determined and a stepwise mechanism of the product formation, involving C-Si bond cleavage by methanol, proposed by the authors.

Above authors[436] studied another photochemical reaction involving silylene with ethene and silene. It was reported that cophotolysis of cyclic organosilanes with phenathroquinone produced silylene inserted products via radical displacement at the silicon atom of the organosilane by the photochemically excited phenanthroquinone molecule.

The photochemistry of dimethylsilylazide has been described by Kuhn and Sander.[437] UV-photolysis (at 248 nm) of matrix isolated dimethylsilylazide produced 1, 1-dimethylsilaanimine and 1-(methylsilyl) methanimine. Upon photolysis at lower wavelength (193 nm), methane and silylisocyanide were formed. The latter is the final product of rearrangement of methylnitrilosilane, which is postulated to be the primary product of photolysing silanimine at 193 nm.

Photochemical reaction of (pi-MeC_5H_5)$Mn(CO_2)L$ with $HMe_2SiXSiMe_2H$ has been studied by Schubert and Grubert[438] and they reported formation of oxidative addition products like Cp′(CO)L.Mn(H)$SiMe_2XSiMe_2H$, by the photochemical reaction of Cp′$Mn(CO)_3$ with 1,2-$(HMe_2Si)_2C_6H_4$ and of Cp′$Mn(CO)_2PMe$ with HMe_2SiXMe_2H (where X = 1,2-C_6H_4, C_2H_4, and O). Except for the combination X = 1,2-C_6H_4 and L = PMe_3, binuclear [Cp′-(CO)(L)Mn(H)$SiMe_2]_2X$ is reported to have additionally formed in the reaction.

Studies on photochemical reactions based on structural features and theoretical quantum molecular orbital energy state calculations, have also been reported and the results verified through spectroscopic and other experiments. Thus, Tachikawa[439] reported photo-reaction mechanism of permethylcyclo-hexasilane via the excited triplet state potential energy surface studied by direct MO-dynamics method. Theoretical calculations indicate that photochemical extrusion of silylene, Me_2Si, from the silicon ring compound is possible on the triplet energy surface as well as on the first excited state surface. Lennartz *et al.*[440] also made similar theoretical calculations for potential energy surface governing the photochemical reactions of silylene with ethene and silylene by *ab-initio* method. These calculations are useful for getting an insight into the photochemical behaviour of silylene (SiH_2) and the reaction of silylene in its first two excited states ($^3B_1$ and $^1B_1$) with ethene and silene. Variations in spectroscopic properties and reactivities in the context of photochemistry of vinyl-disilanes has been discussed as a function of structure and the closely related reactive silenes by Leigh *et al.*[441] through transient spectroscopic data.

Photophysics of permethylated oligo silane chains have been reviewed with a number of references by Raymond[442] in his Ph.D dissertation, while photochemistry of organosilicon compounds covering oligosilanes, polysilanes, and silylenes have been extensively reviewed by Brook[443] recently. Interested readers may consult these two reviews for more detailed information on the subject. The mechanistic aspect of photochemistry (with special attention to non-rigidity of hypervalent silicon compounds and resulting kinetic as well as stereochemical aspects) have been reviewed by Kira.[444] Preparation and photodecomposition of allyl and vinyl containing di and oligo silanes have been reviewed by Semenov *et al.*[445] Due to limitation of space, I will not discuss these here, but readers are referred to the above references for further information on the subject.

(e) Electrosynthesis of silicon compounds

General theoretical principles and applied techniques have been used in the recent past to synthesize organo silicon compounds, which have commercial interest. In these techniques, both permanent electrode and sacrificial anode method is adopted to synthesize the desired product. Some of these examples will be elucidated here.

Graschy *et al.*[446] at Technical University of Graz, Austria, described the process for electrochemical formation of cyclosilanes. In this study, several dichlorosilanes having general formula, R_2SiCl_2, where R = Me, Ph, p-tolyl, were subjected to a systematic study for finding the correlation between bulk substituents and the ring size of the product formed in the respective electrolysis. Furthermore, the influence of a change in anode material and electrolyte system on the electrolysis phenomenon was investigated. Electrolysis of monohydrodichlorosilanes, $RHSiCl_2$ where R = Ph or Me, were not successful in forming cyclic products; instead gave linear polymers due to the high flexibility of the chain. Replacement of the organic group by bulky cyclohexasilanyl substituent did not lead to the expected steric hindrance; instead, electrolysis of $(Si_6Me_{11})HSiCl_2$ yielded bi (undecamethylcyclo-hexasilanyl) as a result of Si-Si bond cleavage and subsequent dimerization of two cyclohexasilanyl groups.

Oshita *et al.*[447] at Hiroshima University, Japan, reported electrosynthesis of silicon containing bridged bithiophenes and the optical, electrochemical, as well as electronic properties of the products. In this study, a series of bithiophene-derivatives, e.g. the compound shown in the figure below (Product I),

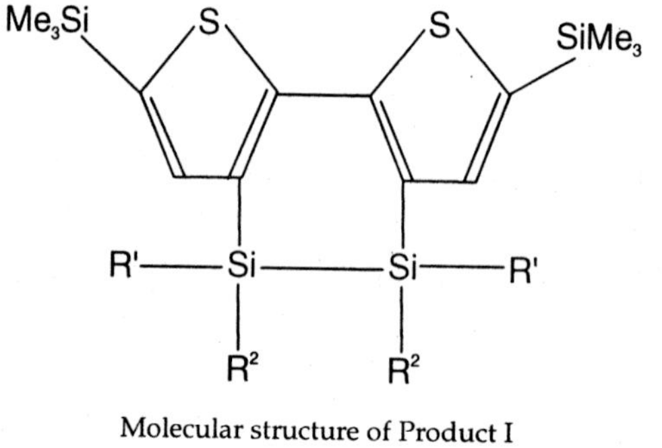

Me_3Si S S $SiMe_3$
R'—Si——Si—R'
R^2 R^2

Molecular structure of Product I

where $R^1 = R^2 = Me$, Et, or $R^1 = Ph$ and $R^2 = Me$, bearing an intramolecular monosilanylene or disilanylene bridge between the β , β´-positions, were synthesized and their properties investigated. UV-spectra and cyclic-voltametry analysis of the silicon bridged bithiophenes indicate that, they have lower lying LUMOs relative to those for bithophene and methylene bridged bithiophenes; probably due to σ^*-π^* interactions between the silicon atoms and bithiophene π-orbitals. The later inference was in good agreement with the results derived by theoretical calculations using simplified model compounds based on RHF/6-31G. The silicon bridged bithiophenes exhibit high electron-transporting properties and triple-layer type electroluminiscent (EL) devices could be fabricated using the silicon bridged bithiophenes, tris (8-quinolinolato) aluminium (III) complex and N,N´-dimtolyl biphenyl-4,4´-diamine (TPD) which works as electron-transporting, emitting and hole-transporting layer, respectively. These triple layer devices emit strong electroluminscence.

From Russia, Martinov and Stepnov,[448] reported electrochemical oxidation of benzyl-silane. In this experiment, electrochemical reduction of some fluorohalocarbons in the presence of trimethylchlorosilane was followed through formation of corresponding fluoroalkyl-chlorosilanes. It was reported that products CF_3SiMe_3, $CFCl_2SiMe_3$, $CF_3CCl_2SiMe_3$, and $(EtO)_2P(O)CF_2SiMe_2$ were obtained with current efficiency between 20 and 50 %. Corresponding voltametric data are also presented in the paper.

At CNRS, France, Constatieux and Picard,[449] reported electrosynthesis of alpha-silylalkylnitrile using sacrificial anode technique. The electrosynthesis was performed in two different ways: (a) use was made of alpha-chloronitriles as the starting material and zinc was used as the material for anode along with THF-HMPA solvent mixture and (b) aluminium was used as the material for anode construction, while nitrile compounds served both as the substrate as well as solvent media. This process was also claimed as an electrochemically efficient process.

From Israel (Bengurion University), Zhang and Becker[450] reported results of controlled potential oxidation and reduction of disilene and tetramethyl-disilene. Results indicate that the main silicon containing products involve only one silicon atom and have the general structure $Mes_2SiX(Y)$, where X and Y are H, Oh, or F.

Electrochemical oxidation of other silicon compounds, like benzylsilanes, have also been reported by Zhuikov.[451] In this study, electrochemical oxidation of substituted trialkylbenzylsilanes were shown to follow the electrochemical exchange (ECE) scheme, including reversible electron transfer reaction and formation of a short-lived radical cation, which can be detected with methoxybenzylsilanes at high temperature sweep rates. Considering three parameter models for the electrochemical reactivity of benzylsilanes, it was found that the inductive effect of alkyl-substiuents at the silicon atom on the rate of fragmentation of the radical cation, was stronger than on the energy of the HOMO. This situation leads to inverse influence on the effective oxidation potential.

From CNRS, France, Rajaonah *et al.*[452] reported electrosynthesis of a new class of functionalized difluoroallylsilanes by an efficient electrochemical silylation route. In this process, silylation of chlorodifluoromethylenolethers, $RCH:C(OEt)CF_2Cl$ where R = Ph or $PhCH_2CH_2$, produced in good yield (94% and 66%) the new functionalized difluoromethylallylsilanes, $RCH:C(OEt)CF_2SiMe_3$ where R is same as above, on a preparative scale. These silanes were found to react as difluoromethyl anion equivalent with an aldehyde (EtCHO) as electrophile, providing functionalized alpha-difluoromethyl alcohols, $RCH:C(OEt)CF_2CH_2(OH)Et_4$.

Kunai *et al.*[453] at Hiroshima University, Japan, reported electrolysis of various halosilanes in the presence of phenylacetylene using a platinum electrode, thus forming Si-sp-C bond; the product from this process was phenylethynylated compounds. Using this technique, silole derivatives were synthesized by electrolysis of bis (phenylethynyl) silanes. On the other hand, coelectrolysis of p-bis (chlorosilyl) benzenes and p-diethynylbenzenes generated polymers composed of an alternating arrangement of p-bis (silanylene) phenylene and p-diethynylene-phenylene units.

Ishifune *et al.*[454] also from Japan (Kinki University), described a general process for electroreductive synthesis of sequence ordered polysilanes using magnesium electrode. In particular, the authors described a technique involving stepwise elongation or growing of Si-Si chain, by electroreductive cross-coupling reaction of chlorohydrosilanes with dichlorooligosilanes. Electroreductive polymerization of the dichlorooligosilanes using magnesium electrodes, is reported to be highly promising for the synthesis of sequence ordered polysilanes. Thus it was inferred that dichlorosilanes are good monomers for the electroreductive synthesis of the polysilanes, where the basic units are ordered in three sequence repeated units.

Besides transfering electron to the compound at electrode surface, studies have also been conducted on dissociative excitation of organosilicon compounds in gaseous state.[455] This type of reactions is more akin to the development of a dissociative process and we will discuss this along other silicon compounds of commercial interest, processes of which have been developed in recent times in a later section (entitled industrial process developments).

(f) Pesticide, fungicide, algicide and bactericide

A number of silicon compounds have been synthesized in recent past which are structurally different from previously synthesized and commercially available pesticides, fungicides, algicides or bactericides. These compounds have the same efficacy, if not better, as its predecessors and seems to be environment friendly. I will quote below some of these compounds, in the following discussion.

Nagase *et al.*[456] at Sagani Chemical Co, Japan, reported preparation of silozanes as bactericides and algicides. They are represented by the general formula, $R'N^+R^2R^3(CH_2)_mSiR^4R^5OSiR^6R^7R^8X^-$, where R′ = hydrocarbon moiety, $R^2 - R^8$ = alkyl groups, X^- = counter ion, and m = 1–6. One such compound, in particular, $C_{18}H_{17}N^+Me_2(CH_2)_3SiMe_2OSiMe_3I^-$, showed *in vitro* MIC activity of 2.5 μg/mL against Bacillus Subtilis.

Yoo *et al.*,[457] reported synthesis and fungicidal activity of 1 (H−1,2,4-triazol-1-yl) alkyl-1 silacyclohexanes (where R = H, Me; X = F; and R = H when X = Ph). These compounds were synthesized by four-step reactions, starting from 1-(chloroalkyl) trichlorosilanes. Their fungicidal powers were tested *in vitro* for 10 different fungi, and *in vivo* assay for four different fungi occuring in rice, barley, tomato, etc. The results were compared with commercially available other fungicidal silicon compounds, namely Flusilazole, and the compound –I, whose structure shown below, especially showed good fungicidal activity with a broad-spectrum close to the results of Flusilazole *in vivo* assay. Synthesis and fungicidal activity of

Si N N R X R = H, Me X = F.

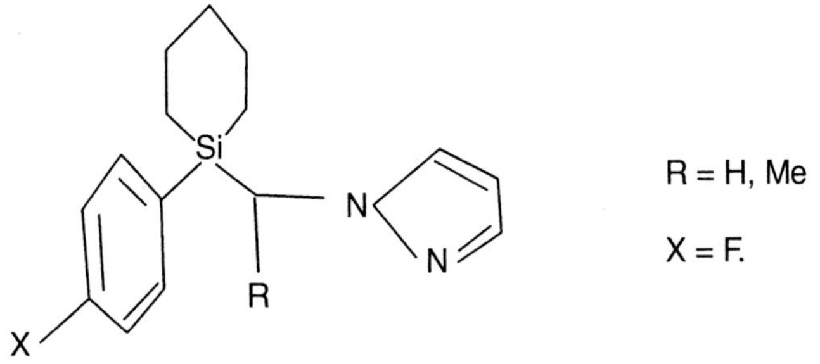

another new class of compounds, S-allyl-O-substituted phenyl-thiophosphates (-onates) was reported recently by He *et al.*[458] from China. They reported isomerization and chlorination of O,O-diallyl-thiophosphate (-onates), such as $(CH_2{:}CHCH_2O)_2P(S)R'$, where R′= PrS, EtS, Me, Et_2N EtO, PrO) with $POCL_3$, producing S-allyl-thiophosphorochloridate (no chloridate), $(CH_2{:}CHCH_2S)P(O)CIR'$ (Product I), and O-allyl-phosphorodichloride $CH_2{:}CHCH_2OP(O)Cl_2$ (Product II). After removal of Product II from Product I under reduced pressure, the crude Product I was further purified by column-chromatography on silica-gel. The purified Product I was then reacted with substituted phenol in chloroform in the presence of triethylamine, giving 18 new title compounds, having structures like $(CH_2{:}CHCH_2S)P(O)R'(OC_6H_4R')$ the Product III. Bioassay results against five common plant disease fungi, like P. Zeac, P. Paricola, R. Solani, A. Solani, and D. Archidicala, showed that the Product III possesses some excellent fungicidal activity at a mere 0.005% concentration.

The same group of workers in China, reported[459] synthesis and insecticidal activity of silicon-containing asymmetric thiophosphates and their carbon analogs. In particular, two series of silicon-containing asymmetric thiophosphates (ETO)(PhO)P(O)SCH$_2$SiMeR^{1}R^2 where R^1,R^2 = alkyl, (un)substituted Ph, and the other class (R^{3}O)(PrS)P(O)OC$_6$H$_4$EtMe$_3$ where R^3 = Me, Et, Pr, Bu, i-C$_5$H$_{11}$, or PhCH$_2$, were synthesized for finding out their insecticidal activity. The carbon analogs of one series of silicon compounds were also prepared and the same activity compared. The paper discusses these results along with the $_1$H, ^{31}P NMR spectra of the compounds.

Barnes and Fu,[460] on the other hand, studied insecticidal and acaricidal strength of fluorosilicons, namely difluorovinylsilane. These categories of compounds having general formula ArSiRR′CF : CFCHR^{2}Ar′ (compound I, structure shown below) where Ar = halo-C$_{1-4}$ alkyl, halo-alkyl, or alkoxy (un)substituted, Ph, etc., and R,R′= H, C$_{1-4}$ alkyl, C$_{3-5}$ cycloalkyl; R^2 = H, cyano, C$_{1-4}$ alkyl or haloalkyl, etc., when Ar′ = (un)substituted phenoxyphenyl, etc., were synthesized and found to have useful property as insecticide or acaricide.

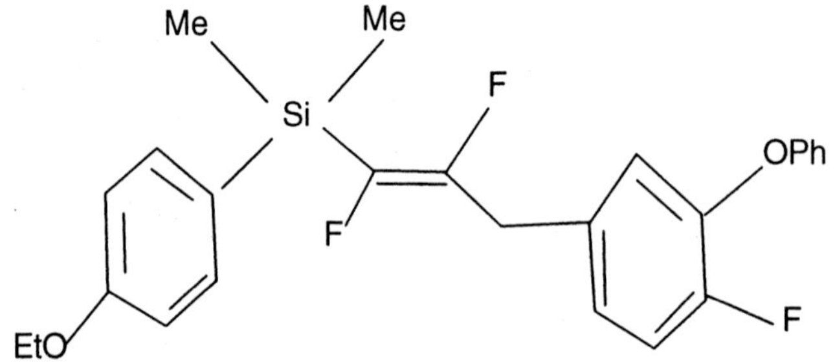

Compound-I

Thus, p-EtOC$_6$H$_4$SiMe$_2$CF:CFH was treated with BuLi and the product was later reacted with alpha-bromo-4-fluoro-3-phenoxy toluene in presence of (Ph$_3$P)$_4$Pd and zinc-chloride to produce the difluorovinylsilane (Compound II). This latter compound, at mere 100 ppm concentration showed 100 % insecticidal effect for Spocloptera Eridania and Heliothis Vivenseens insects.

(g) Silicon compounds for biological and analytical applications

Several new silicon compounds which can be used for medicinal purpose (both internal and external application) have been reported in recent times. Some of these compounds are currently under actual trial on human bodies and expected to be commercially produced soon.

One such new compound reported by Pikies and Ernst[461] from Technical University, Gdansk, Poland, is a new class of muscarinic-agent. Muscarinic compounds are those, ingestion of which affects both muscular and mucosal tissues in human body, thereby causing excess sweat, saliva discharge, bronchial secretion, abdominal colic, etc. Structurally, antimuscarinic agents are the silicon analog of the conventional drug biperiden. Structure of this compound a racemic-(exo-bicyclo(2,2,1)hept-5-en-2yl)phenyl(2-piperidinoethyl)silanol is as follows:

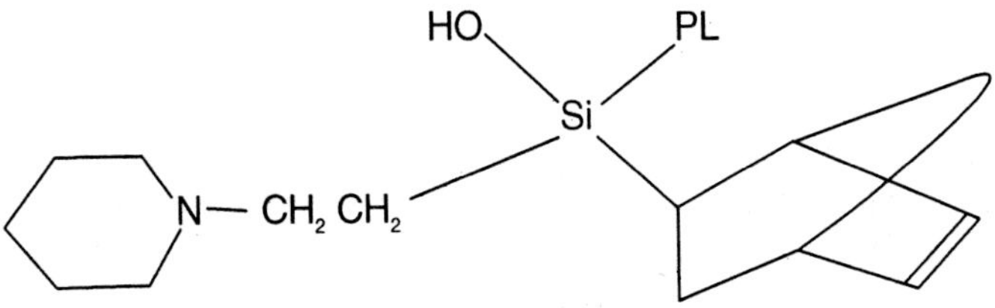

(racemic mixture)

The compound was prepared in seven steps; the first step consists in the hydrosilylation of 5-exo-bromobicyclo (2,2,1) hept-2-ene with dichloro (phenyl) silane, in the presence of chloroplatinic-acid. The product of the last synthetic step is a mixture of four endo- and four exo-diastereomers of (bicyclo(2,2,1)hept-5-en-2yl)phenyl(2-peperidino-ethyl)silanol, and the (tricyclo ($2.2.1.0^{1.5}$) hept-1-yl) silanol derivative. Eight times crystallization produced pure silabiperiden (racemate) in an overall yield of 5.3%, which represents a more than 30-fold improvement over the earlier synthesis.

From Roche Laboratory, Switzerland, Huber[462] reported a group of silanyl-triazines which can be used as light screening compound. In particular, S-triazines, whose molecular structure is shown below, was synthesized. In the structure, W^1, W^2, W^3 each independently signifies C_1–C_{20} alkyl or a group of Sp.Sil structure (Sp = a spacer group & Sil = silane, oligosiloxane, or polysiloxane moiety); X^1, X^2, X^3 each independently signifies O or NH, with the provisio that at least one of W^1, W^2, and W^3 signifies Sp.Sil group. Thus, a divinyltetramethyldisiloxane-platinum complex catalyzed hydrosilylation reaction of 3-butene-1-ol with pentamethyl disiloxane gave 4-(1,1,3,3,3-pentamethyl-disiloxanyl)-1-butanol, which on $Ti(OPr)_4$ mediated reaction with Et_4-aminobenzoate gave 4-aminobenzoic acid 4-(1,1,3,3,3-pentamethyl-disiloxanyl) butylester. Reaction of the latter

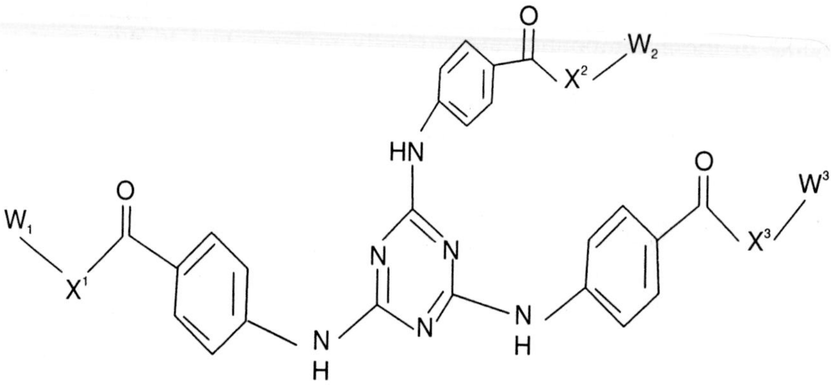

Compound-I

compound with cyanuric-acid-chloride gave the compound I shown above, where X^1, X^2, X^3 = O and W^1, W^2, W^3 = $CH_2(CH_2)_3SiMe_2OSiMe_3$. This compound can be used as UV-B filter in light screening (sunlight) compositions, especially for preparation of cosmetics used for protecting the human skin from harmful cancer producing ultraviolet radiation.

Wright and Busch[463] have described a compound 2,6-di-alkyl-4-silyl-phenols along with its structure, which can be used for inhibiting vascular cell adhesion and treating chronic inflammatory disease. 14 anti-inflammatory silyl-phenols (see molecular structure below) were prepared by the above workers and by reaction of 1, 2, 6-$HOR_2C_6H_2XH$-4 with $ClCH_2SiMe_2R^1$,

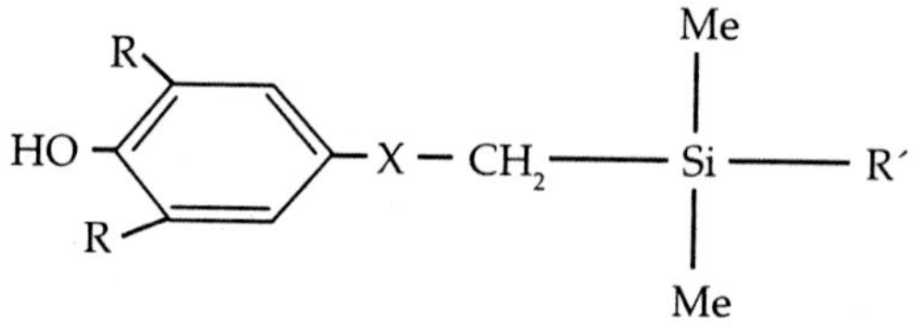

having yield in the range of 19-64 %. In the above structure, R^1 = Ph, Me, p-ClC_6H_4, p-FC_6H_4, o- and p-anisyl, 2,5-$(MeO)_2$-2, 3-$(MeO)_2C_6H_3$, p-$Me_3CC_6H_4$, Ph.CH_2; R = Me_3C, Me; and X = O, S.

Lukevies *et al.*[464] from Latvia, reported silyl modification of biologically active silicon-compounds. Thus, the derivatives of (1,2,3,4-tetrahydro-1-quinolyl)-, (1, 2, 3, 4-tetrahydro-2-isoquinolyl)-, and (1,2,3,4-tetrahydrosila-2-isoquinolyl) acetic acid, which are structural analogs of glycine, were synthesized. The psychotropic activity and acute toxicity of the compounds have been studied and mentioned in the paper.

Besides above, biocatalytic kinetic resolution of hydroperoxyvinyl-silanes by horseradish-peroxidase as well as by lipase, reported by Adam *et al.*,[465] and synthesis as well as properties of novel silanediol-protease enzyme-inhibitors reported by Chen.[466]

(h) New process routes for industrial compounds

These studies are mostly aimed at synthesizing intermediate or final product related to industrial processes. Attention has also been paid to purify the final product, in order to arrive at a better defined property and higher efficacy of the final products. Processes of this kind that have been reported, being large in number, and due to limitation of space and scope of this book, I will mention here only the important title compounds or the process route. Readers interested in further details may consult the full paper whose reference is quoted alongwith:

1. A process for preparation of tertiary alkylsilanes.[465]
2. Preparation of silicones having siloxy groups.[466]
3. Process for the preparation of propyltrichlorosilane.[467]
4. Improved process for obtaining organosilanes using Lewis acids.[468]
5. A new route to silicon-alkoxides from silica.[469]
6. Stabilization of purified trimethoxysilane by starring in metal container.[470]
7. Process for converting polymeric silicon containing compounds to monosilanes.[471]
8. Continuous process for preparing aminopropyl-trialkoxysilanes.[472]
9. Preparation of alkoxysilyl containing silatranes as coupling agents or as adhesion promoters.[473]
10. Silsesquioxanes: a key intermediate in the building of molecular composite materials.[474]
11. Fluoroalkyl containing organosilicon compounds and their use.[475]
12. Process for the production of octaphenylcyclotetrasiloxane.[476]
13. Preparation of trimethyl-chlorosilane.[477]
14. Cyclic-siloxanes and their use as wetting aids and foam-stabilizer.[478]
15. Organosilicon chemistry (from molecule to material).[479]
16. Method for producing organosilanes using redistribution reaction catalyzed by alumina based materials.[480]
17. Preparation of organosilanes by redistribution reaction catalyzed by Lewis acid using a catalyst-inhibitor after redistribution.[481]
18. Polyfunctional organosilane coated silica.[482]
19. Managing a technical revolution: the switch from trichlorosilane to trimethoxysilane based processes.[483]
20. Continuous organomagnesium synthesis of phenyltrichlorosilane.[484]
21. The direct production of methylchlorosilane.[485]
22. Synthesis and application of omega-epoxy functionalized alkoxysilanes.[486]

23. Thermal stability (to-75°C) of 1-lithio-2-trimethylsiloxyethylene in diethyl-ether solution.[487]
24. Procedure for production of acyloxyalkoxysilanes.[488]
25. Preparation and synthetic utility of oxasilacyclopentane-acetals derived from silaranes.[489]
26. Preparation of organosilicon compounds as surface treatment agents and resin additives.[490,491]
27. Process for preparation of alkynylsilanes by catalytic coupling reactions.[492]
28. High purity branched phenylsiloxane fluids.[493]
29. Preparation of organosilicon compounds.[494]
30. Synthesis of N-(trialkylsilyl) morpholins and their use as hydrophobic layer on silica particles.[495]
31. Cyclo-organosilyl derivative for determination of alcohols and carboxylic acid by gas chromatography (G.C.) mass-spectra.[496]
32. Preparation of allytrialkoxysilanes as resin modifiers or coupling agents.[497]
33. Preparation of acetylene group containing silanes as materials for heat-resistant polysilanes.[498]
34. Preparation of 1,4-dioxanyl group containing dihydroxysilane as an intermediate for hydrophillic-silicones.[499]
35. Preparation of 1,4-dioxanyl group containing dialkoxysilanes as an intermediate for hydrophillic-silicones.[500]
36. Purification of alkoxysilanes.[501]
37. Process for purification of tetraethoxysilane using chromatographic separation column.[502]
38. Purification of dimethoxymethylsilane by azeotropic removal of methanol.[503]
39. Purification of dimethoxymethylsilane by removal of methanol.[504]
40. Purification of organosilicones of group IIIa and Va by complexation.[505]
41. Preparation of silicon-peroxide compounds.[506]
42. Model compound for diphenylsiloxane polymer (octaphenyltetrasiloxane-1,7 diol and its organo-tin derivatives).[507]
43. Method for synthesis of chlorosilanes.[508]
44. Preparation of alkoxysilanes from silicon and alcohols.[509]
45. Preparation of alkoxysilanes from silicon and alcohols.[510]
46. Preparation of alkoxysilanes by recycling alcohols.[511]
47. One vessel synthesis of N-(methyldihalosilylmethyl) amides and lactam.[512]
48. One step preparation of alkynylsilanes from halosilanes and 1-alkynes.[513]
49. One step preparation of 1-alkynylsilanes from halosilanes and alkynylcopper.[514]

50. Use of surface active additives in the direct synthesis of trialkoxysilanes.[515]
51. Preparation of organic-silicon compounds containing sulphonic acid groups as modifiers and coupling agents.[516]
52. Preparation of organic-silicon compounds as coupling agents.[517]
53. One step process for converting high boiling residue from direct processes to monosilanes.[518]
54. First step in the direct synthesis of methylchlorosilane.[519]
55. Process for preparing N or S containing organosilanes.[520]
56. Continuous organomagnesium synthesis from mixtures of ethylethoxysilanes with methyl (thienyl) or halo-organo-ethoxysilanes.[521]
57. Preparation of alkoxysilanes containing little quantity of halogens.[522]

- Besides the above, a large number of hetroatomic compounds, silylation reactions and basic preparative reactions have been reported in the past few years. Fundamental reaction studies involved stereoselective and regioselective synthesis as well as various catalyzed reactions. Since the number of such publications is large, we will discuss each of these category in short and corresponding references will be cited for interested readers for further consultation.

(i) Heteroatomic compounds

Some of the recent publications, showing the trend in research in this direction, are listed under references 523-539. Heteroatoms that have been included under this subheading includes lithium, phosphorus, boron, silver, sulphur, fluorine and transition-metal atoms. Hetero pi-systems include compounds like-CHNSi, CNSI, $C_2H_4Si_2$, C_3H_4Si, etc. Presence of heteroatom shift the electrondensity in the silicon polymeric chain or in the molecules, thereby adding selective reaction site with electrophilic or nucleophilic reagents.

(ii) Catalyzed reactions

Recent literature available under this category are shown in the references 540-577. These reactions include all three phases like solid, liquid and gas phase reactions. Catalysts tried mostly are the following metals or their complexes rhuthenium-complex, copper Schiff-base complexes, iridium, platinum, titanocene, copper salts, transition metal complexes, solid acids, Lewis acids and bases, thiols, palladium and nickel metal catalysts, ruthenium, and silver. Influence of external factors on reaction kinetics, like ultrasound, laser light, and electron beam, have also been studied and included in the above references.

(iii) Silylation reactions

Silylation reactions include the following radicals:

Silyl	$-SiH_3$	Siloxy	$-OSiH_3$
Silylene	$=SiH_2$	Silylthio	$=S.SiH_3$
Silylidyne	$-SiH$	Silylamino	$-NHSiH_3$

Most of these references have been cited earlier (Section 5.4), and additional related reactions involving migration of silyl group (inter-and intramolecular migration) are included in the next subsection 'rearrangement reactions'.

(iv) Rearrangement reactions

Recent publications of rearrangement reactions, like migration of clusters, Sila-Pummerer Rearrangement reactions, dyotropic rearrangements, thermal-rearrangements, rearrangement through addition/elimination reactions, ring-inversions, isomerizations, cage rearrangement (in silesesqui derivatives), and cross-coupling reactions, are tabulated from reference 578-599. Interested readers are referred to the above papers for further details.

(v) Fundamental preparative reactions

Some of the interesting preparative reactions currently being published are shown from references 600-817. While some of these preparative reactions have importance for synthesizing intermediates, the rest is aimed at fundamental studies like following reaction path, studying steric effect of bulky groups, various substituent effect studies, kinetic studies, structure determination, studies on ring closure mechanism and ring enlargement paths, cycloaddition reaction studies, enantio-selective reductions, stabilization studies, studies on free radical mechanisms, gas phase disproportionation and coproportionation reactions, zwitterionic compounds and spirocyclic compounds, polyhedral silicon compounds, synthesis of asymmetric molecules, newer variety of nucleophilic and electrophilic addition reactions, stereoselective addition reactions, and synthesis reactions occurring through flash vacuum thermolysis, etc. Results of these studies have benefits of both giving insight into near synthetic routes and deriving new compounds with special properties that will be of interest in the near future.

(vi) Combinatorial chemistry of organo-silicon compounds

This is relatively a newer area and slowly emerging as a powerful tool for fast processing and designing synthetic experiments involving a large number of possible silicon compounds. Combinatorial chemistry has the special advantage of being able to synthesize a large number of compounds based on general organic chemistry principles (which also includes polymeric and biochemical compounds) from available informations through high power computing technique and thus providing path for selecting a small number of laboratory studies and consequently eliminating the need for a large number of hit and trial piecemeal experiments, thereby saving time and energy. Accordingly, combinatorial chemistry provides advance prediction of some useful properties of new compounds (as is possible from a large number of silicon compounds) based on their structure. Combinatorial chemistry of a large molecle, like fullerene, has also been tested recently by Dr S.R. Wilson at Sphere Biosystem Inc., New Jersey (USA).

Processing step in combinatorial chemistry starts with defining various strategies and processes for the rapid synthesis of large, organized collection of compounds called 'libraries'. Next step is based on this file finding out active compounds. This combined method has already helped producing a collection of molecularly diverse compounds that can be used for rapid screening for biological activity. In the beginning, this new branch of chemistry started as application to drug discovery process, which later expanded its application to analytical and synthetic chemistry. Parallel organic synthesis has been performed successfully in combinatorial chemistry using Parke-Davis diversomer technology based on split-mix method, pioneered by Furuka *et al.*[818] Excellent books are now available[819] in the market, which readers can consult for further detailed procedure of this new chemistry. Here I will cite some of its recent applications in the field of silicon-chemistry.

Hone and Bexter[820] at Oxford Asymmetry Ltd, UK, applied combinatorial chemistry in the synthesis of organosilicon compound for particular application. Compounds like p-$RC_6H_4SiMe_2CH_2CH_2COOH$ and p-$RC_6H_4SiMe_2CH_2CH_2CONHQ$, where R = Br (for Product-I), Me_3Sn, 1-naphthyl, p-$MeOC_6H_4CONH$, p-tolyl, etc., and Q = polystyrene resin, were prepard by use of the above process. Product-I (above) was prepared in four selected steps from p-dibromobenzene using known standard methods.

5.6 GENERAL SPECIFICATION OF RAW MATERIAL, CATALYST AND PROMOTER (VARIOUS DESIGN OF REACTORS)

Specification of raw materials

Metallic silicon is the primary raw material for the synthesis of silicon monomers as well as silicon halides, which in-turn serves as the raw material for further processing to finished material. Earlier, silicon having a purity of around 98 % was sufficient for this purpose, but later higher purity (purity exceeding 99 %) became the norm for better yield and higher purity as demanded by the market. Minute impurities in these silicons were found to significantly affect the course of the reaction and thus the purity of the desired product. Some of the silicons are also difficult to react and thus needs activation. General activation steps involve pretreating the silicon with hydrofluoric-acid or other highly corrosive acids, in order to etch away the refractive inert oxide coating on the outer surface of silicon particles. Other methods of activation includes treating the silicon at 1000°C in a stream of nitrogen or preferably hydrogen. Particle-sizes of the silicon is another important parameter. Finer the particle, more is the available surface-area and faster is the rate of reaction. But finer particle attribute more towards carryover in the ensuing gas stream of unreacted particles and imposition of additional gadgets like recirculation system. This later condition is particularly true for fluidized or entrained-bed reactors. Much work, accordingly has been carried out to optimize this size range, and generally silicon of 30-350 μm and copper (catalyst) 5-30 μm suffice for the purpose.

Similarly, purity of the catalyst is also important in production of silicon compounds. Electrolytic grade copper is further purified to meet the above objective. Similar purity level needs to be maintanied for other catalysts as well (e.g. with palladium, rhodium, platinum, rhuthenium, Lewis acids, aluminium-trichloride, solid acids and copper complexes) used in down stream industries. The optimum amount of catalyst to be used for silicon production, depends on the particular reaction being followed; but in general, 5-10 % of the metal (e.g. copper or silver) is used in most commercial productions. Promoters like zinc (either as pure metal or its compounds like oxide or chloride) are also used to help the reaction proceed selectively at a desired rate, along with the catalysts.

In stirred or moving beds (e.g. entrained or fluidized-bed or circulating fluidized-bed reactors), the raw materials automatically gets thoroughly mixed in the course of the reaction, but in the case of stationary beds (like fixed bed reactors), it may be necessary to alloy the above two raw materials before putting them into the reactor. Sintering at 1000-1100 °C in a stream of hydrogen or 10% nitrogen, serves the above purpose. Silicon powders in some instances, also have been superficially alloyed with catalyst metal salts (e.g. copper(I)chloride) at 200-400 °C.

The by-product, such as hydrochloric acid (which can be converted to chloromethane by reaction with methanol), or the chloromethane produced in methanolysis reactions are recycled back into the process after necessary purification, in order to avoid environmetal pollution and help in economy.

Promoters

Maintaining purity of silicon and the catalyst do not guarantee desired yield or reaction rate in such reactions. It may be necessary to supplement the shortfall by incorporating a third element known as 'promoters', added at a low rate of 0.05-0.5 %. These promoters are capable of increasing both the activity of the catalyst-silicon mixture and selectivity of the synthesis process. Materials generally used for this purpose are antimony, cadmium, aluminium, zinc, tin, phosphorus or a combination of these. It is important to control promoter concentration in a narrow range, beyond which it may inhibit the reaction.

Design of reaction vessel

In commercial production, if the reactants are solid and gaseous materials, generally the reactions are carried out in fluidized-bed reactor (Fig. 9). Fluidized bed reactors have the advantage of being compact (low capital cost) and rate of reaction very fast. One example under this category, is the production of trichlorosilane from silicon metal and gaseous hydrochloric acid (the process also produces silicon-tetrachloride, and dichlorosilane), as well as production of chloromethylsilanes. The solid raw materials, silicon and the metal catalyst, are brought to a fluid state by purging dry hydrochloric acid gas through it.

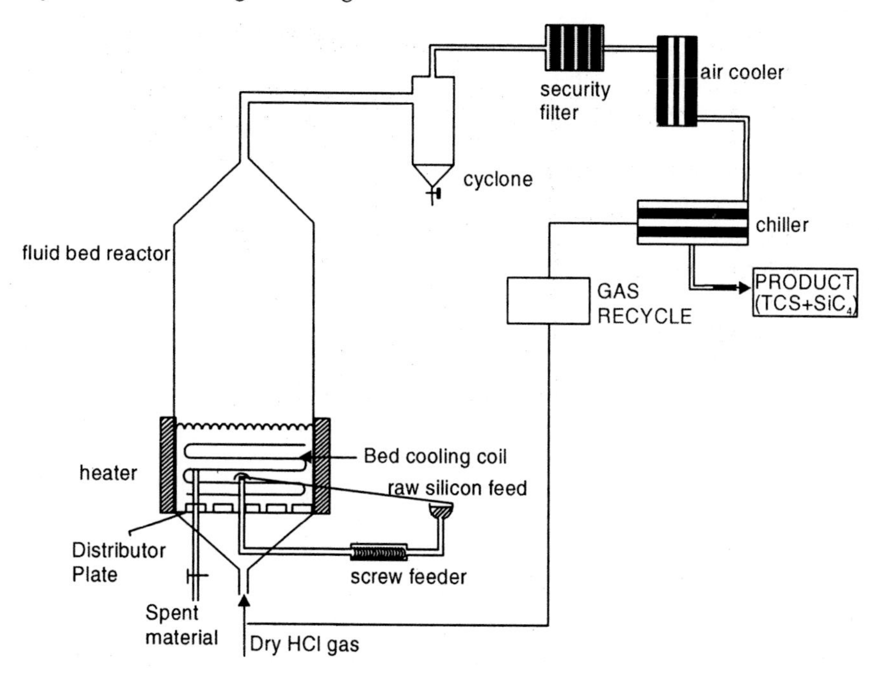

Fig. 9. Schematic diagram of a fluid-bed reactor system.

Simultaneously heat is being applied from outside (at the start of the reaction) on the surface of the reactor. Since the reaction is exothermic, once initiated it proceeds in its own heat and accordingly external heat is cut-off as soon as the exothermic heat becomes available to the reaction mixture. In case of excess heat generation, which may be detrimental to desired product formation, arrangement for cooling through in-bed cooling coil is necessary. For chloromethyl-silane production, the solid raw materials (silicon and the catalyst), is fluidized by the incoming chloromethane gas at about 300°C and 2-5 bar pressure. In both the processes, product gas carries solid particles which first need to be separated by a conventional dry cyclone and then through a fine fabric filter, and finally cooled and condensed to liquid product. In trichlorosilane production, the by-product silicon-tetrachloride and in chloromethane production the by-product silane, is separated and recycled back into the system. Depletion of fresh raw material in the bed is continuously made up by a screw feeding system. In fluidized bed reaction, 90-98% of the silicon is consumed in the production process.

Material for reactor construction is important, if the reacting gas is acidic as in the above case. For dry hydrochloric acid gas and for methylchlorosilane (particularly CH_3Cl), steel is the material of choice. Methylchlorosilane reacts with zinc, tin, magnesium, and aluminium at elevated temperature. Thus they are avoided in the alloy composition of the metal of construction for the reactor. Other materials used in different sections of the plant (flexible joints, pipe, bends, valves, etc.) are enamel or teflon coated materials. However, corrosion problems can be avoided if halogen-free reactants are used, like treatment of silicon with dimethylether producing halogen-free methoxymethylsilanes. Such[822] materials which can also serve as starting material for production of silicones, however, need drastic reaction, and thus are unlikely to become economic alternative to conventional direct synthetic mentioned above.

Fluid-bed reactor output gas is generally a mix of products, and distribution of the products are mainly governed by the operating variables. Thus the same reactor and reactants can be used in producing one product in preference to the other.

If both the reactants in the production process are liquid, the preferred design is the stirred reactor. One such case is the production of tetraethoxysilane (Fig. 10) from the reaction between silicon-tetrachloride and ethanol:

$$SiCl_4 + 4\ C_2H_5OH = Si(OC_2H_5)_4 + 4\ HCl$$

The acid produced (HCl) in the process, needs to be removed immediately, in order to prevent self-propagated by-product reaction, which results in undesirable polymer formation.[823,824] In practice, this is done by blowing dry air or nitrogen through the mixture, or by applying vacuum, use of refluxing solvents like hydrocarbons in the stirred reactor. Amine can also be used as the base-acceptor, but this is not feasible in commercial practice.

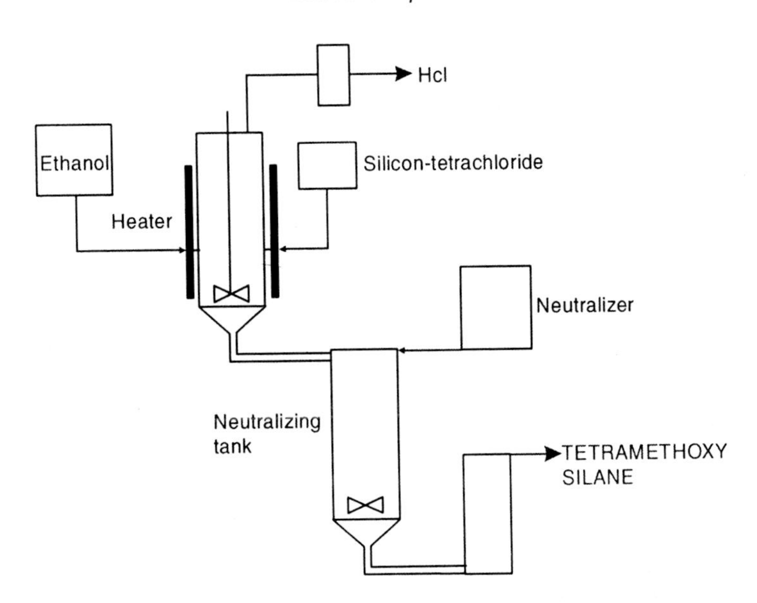

Fig. 10. Flow chart for tetraethoxysilane.

The reactor used for the above reaction, is provided with external heater for initial heating, as esterification process is endothermic (evaporation of HCl cools down the mass). Tertiary alkoxide is not made by this process, however. The other commercial route for alkoxysilane preparation is transesterification:

$$Si(OR)_4 + 4\ R'OH \xrightarrow{cat} Si(OR')_4 + 4\ ROH$$

Since it is an equilibrium reaction, it is applicable when the alcohol to be esterified has high boiling point and product alcohol can be removed by distillation.

For production of silicon polymers (Fig. 11), the following type of reactors are being commonly used: (a) fixed-bed catalyst reactor, (b) cascaded stirred reactor, and (c) continuous screw extruder type reactor.

Polymerization of low molecular weight siloxanes, in small-scale production, can be carried out in a stirred vessel (up to 15 ton/batch) reactor. But for large scale production, continuous plants, as shown in Fig. 11, are used.

Hydrolysis reaction of dichlorodimethylsilane, on the other hand, is carried out in a continuous circulating type liquid phase reactor (Fig. 12) with about 25 % HCl or in gas phase at about 100 °C. Liquid phase hydrolysis gives cyclic and linear oligomeric dimethyl-siloxanes in 1:1 to 1:2 ratio.

A mechanically stirred fluidized-bed reactor (Fig. 13) has also been used to produce chlorosilane and silane by reacting gaseous methylchloride

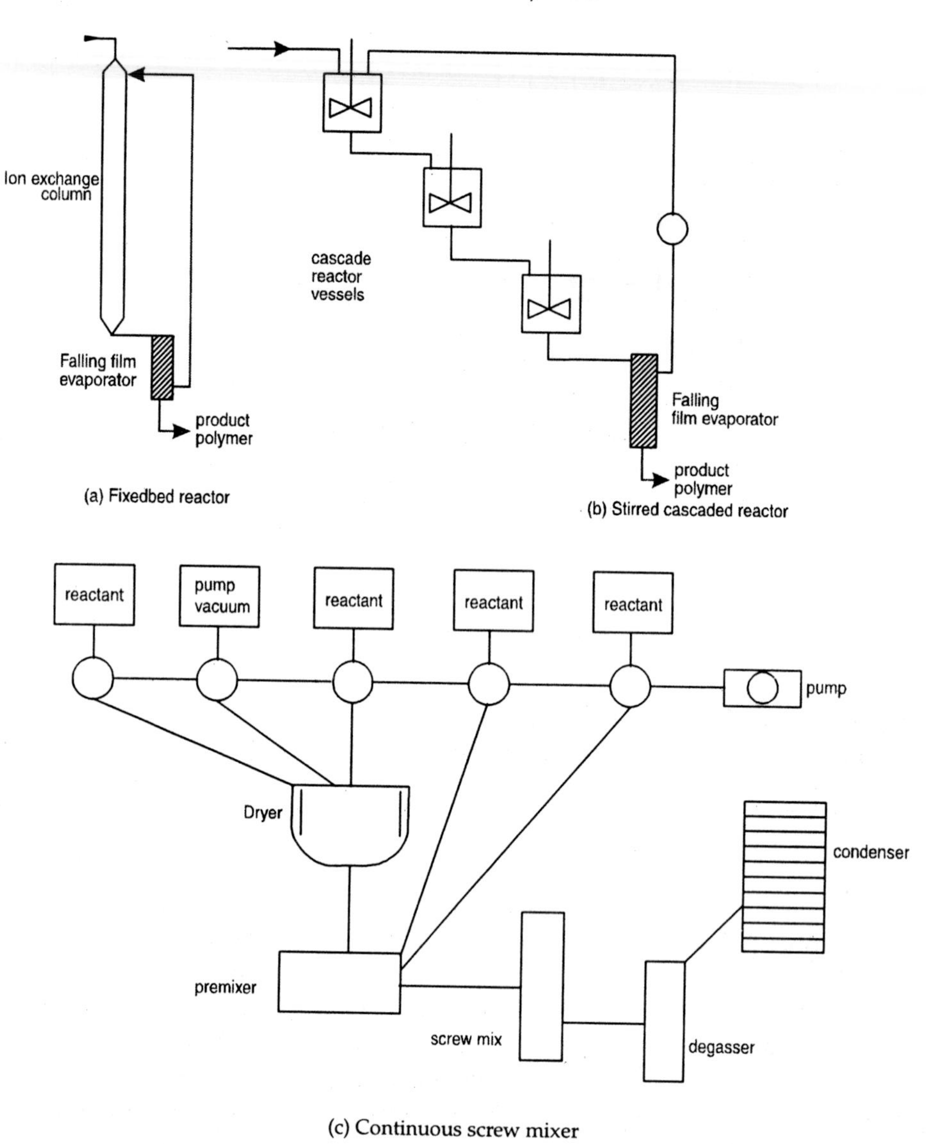

Fig. 11. Reactors for production of silicon polymer.

with silicon powder-copper catalyst.[825] Stirring causes up and down motion of the silicon powder in the vertical reactor and is caused by a ribbon-blender shaft. Ethylchlorosilane has also been prepared by reaction of ethylchloride,[826] and by ethylene and HCl with silicon powder.[827]

Other reactors that can be used are bubbling-bed (e.g. for the Grignard synthesis), when the reactor is packed with magnesium rings, and liquid

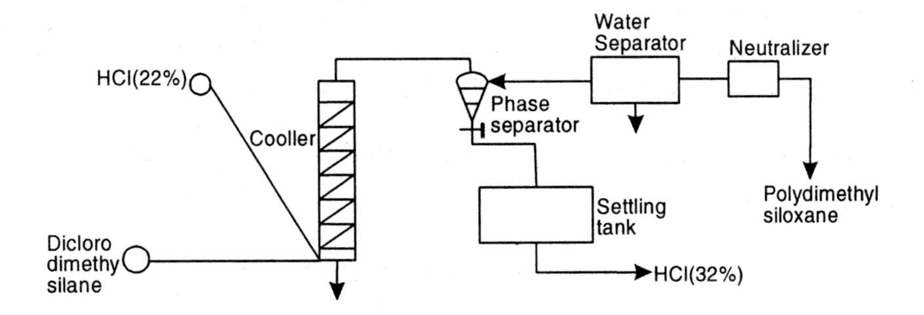

Fig. 12. Continuous hydrolysis of dichlorodimethylsilane in external recycle reactor.

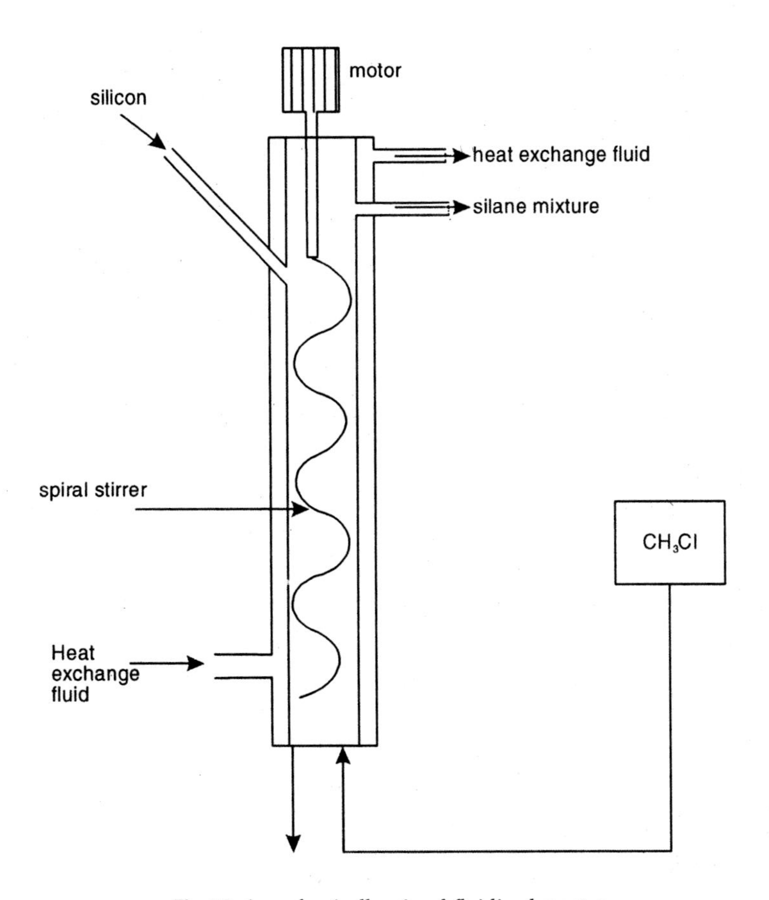

Fig. 13. A mechanically stirred fluidized reactor.

as well as gaseous reactants flow through the reactor in counter-current fashion.

5.7 FUNDAMENTAL PREPARATIVE REACTIONS IN SILICON CHEMISTRY

A good number of text books are available in the market on standard reaction paths for synthesizing silicon-compounds. Since plenty of information is available about these synthesis technique using various starting materials (analogous to carbon-hydrogen organic synthesis), I would like to discuss some of the possibilities of synthesizing most of the compounds starting with the familiar two-silicon products—trichlorosilane and silicon-tetrachloride, production of which has been discussed in detail in the previous chapters concerning production of high purity silicon metal. One of the reasons to emphasize on these two basic material sources being that integration of plant productions for various final product synthesis, is becoming a key issue nowadays, when silicon-compounds are facing stiff competition in a competitive market. This point will be further clear in the next chapter (Chapter 6), when we discuss economic aspects of silicon compounds. On the other hand, Ritscher,[483] has emphasized this integration through trimethoxysilane rather than trichlorosilane. Ritscher[483] has set up an integrated demonstration plant in his work place, WITCO Corporation, Friendly, West-Vergenia 26146 (USA). However, the later idea being still in demonstration stage, we will stick to more conventional products like trichlorosilane and silicon-tetrachloride.

Basically all reactions in silicon chemistry follow three routes—direct synthesis (involving direct reaction between silicon and organic halide), organometallic synthesis (where an organometallic compound transfers the organic radical to the silicon atom), and reaction between silane and hydrocarbons.

Halosilanes ($SiCl_4$ and $HSiCl_3$) are very reactive with variety of protic solvents and thus final products can be made by direct synthesis with these two starting materials. For example, their reactions with alcohols and amines yield the following products:

$$SiCl_4 + 4\ CH_3CH_2OH \rightarrow Si(OCH_2CH_3)_4 + 4\ HCl$$
$$SiCl_4 + HN(CH_3)_2\ \text{excess} \rightarrow Si[N(CH_3)_2]_4 + 4\ HN(CH_3)_2HCl$$

These are reversible reactions and the forward reaction is facilitated by the removal of HCl being formed (e.g. by adding tertiary amine or excess amine reactant, which generally precipitates as amine-hydrochloride.

The monomeric alkylsilanes (used for making silicone fluids and resins) can be prepared by reacting trichlorosilane with alkenes, as:

$$HSiCl_3 + RCH{=}CH_2 \rightarrow RCH_2CH_2SiCl_3$$

While the above two methods show the route for attaching C–C and C = C hydrocarbon directly to silicon, the following reaction shows the method of attaching C≡C to silicon:[828]

$$HSiCl_3 + LiC{\equiv}CLi \rightarrow (\underset{\displaystyle Cl}{HSi}{-}C{\equiv}C)_n + 2LiCl$$

or, in general

$$Cl{-}\underset{\displaystyle R}{\overset{\displaystyle R}{Si}}{-}Cl + LiC{\equiv}CLi \rightarrow (\underset{\displaystyle R}{\overset{\displaystyle R}{Si}}{-}C{\equiv}C)_n + 2LiCl$$

The product LiCl is washed away from the polymer. These polymers are known as 'silylene-acetylene' polymer or copolymer. Yield of the polymer is 100% when R and R´ are different. However, the polymer yield drops due to formation of a small amount of cyclic and linear low molecular weight polymers (oligomers). The organometallic compound, LiC≡CLi, is prepared in the same reaction pot by reaction between trichloroethylene with butyllithium at –78°C, as:

$$3BuLi + (Cl)_2C{=}C(Cl)_2 \rightarrow LiC{\equiv}CLi + LiCl, BuCl, BuH$$

Silylene-acetylene polymers are soluble in a number of organic solvents, and can be (especially those with large cyclic R and R´ group) can be melt spun at 200°C into uniform 10 μm diameter fiber. These fibers are further cross-linked by UV-light from a low pressure mercury vapour lamp (wavelength 254 nm) in an inert gas atmosphere. Cross-linking prevents the fiber from remelting on pyrolysis. Pyrolysis of these fibers at 1200°C under helium/argon atmosphere yields ceramic fibers. Example of some of the melt-spinnable polymer under this category are:

$$\overset{\displaystyle 2Chex}{(Si{-}C{\equiv}C)_n},\quad \underset{\displaystyle Me}{\overset{\displaystyle chex}{(Si{-}C{\equiv}C)_n}},\quad \overset{\displaystyle 2Ph}{(Si{-}C{\equiv}C)_n},\quad \underset{\displaystyle Me}{\overset{\displaystyle Ph}{(SiC{\equiv}C)_n}},\quad \underset{\displaystyle Me}{\overset{\displaystyle Benz}{(Si{-}C{\equiv}C)_n}}$$

Compounds which are not melt spinnable under the same category of compounds are:

$$\overset{\displaystyle Me_2Me_2}{(Si{-}Si{-}C{\equiv}C)_n},\quad \overset{\displaystyle Me_2}{(Si{-}C{\equiv}C)_n},\quad \overset{\displaystyle Et_2}{(Si{-}C{\equiv}C)_n},\quad (2isoC_3H_7{-}Si{-}C{\equiv}C)_n,$$

$$(Bu_2Si{-}C{\equiv}C)_n,\quad (Hex_2Si{-}CC)_n,\quad \underset{\displaystyle Me}{\overset{\displaystyle Hex}{(Si{-}C{\equiv}C)_n}}.$$

where chex = C_6H_{11} cyclic, Ph = C_6H_5, Et = C_2H_5, Hex = $n{-}C_6H_{13}$, Benz = C_7H_7. Pyrolysis of these products results in weight loss, which increases with the mass of aliphatic side group. Polymers with aromatic side groups loose less weight on pyrolysis than its aliphatic counterpart.

Dichlorosilane (H_2SiCl_2) can also be condensed with sodium to produce polysilane, which is then converted to polycarbosilane by an elaborate process. Polycarbosilane polymer fibers can be drawn and air cured. Pyrolysis of these polymeric fibers in an inert atmosphere also results in ceramic fiber.

The direct synthetic route of Rochow[829] also provides the means for attaching hydrocarbon groups directly to silicon. This process involves reaction of organic halide with silicon or silicon alloy. Thus the reaction can be represented by an ideal equation:

$$2\ RX + Si = R_2SiX_2$$

This reaction generally accompanies other side reactions, such as:

$$3\ RX + Si = RSiX_3 + 2\ R. \ ;\ 3\ RX + Si = R_3SiX + X_2 \ ;\ 2\ X_2 + Si = SiX_4.$$

The first equation above, free radical generation reaction, can further give way to organo-H-halosilanes, such as $RSiHX_2$ and R_2SiHX.

In the above reactions, if instead of RX, a mixture of olefins and HCl is passed over metallic silicon, then the following reaction occurs.[830]

$$3\ HCl + Si = HSiCl_3 + H_2, \text{ and then } HSiCl_3 + CH_2{=}CH_2 = C_2H_5SiCl_3$$

Reaction of lower aliphatic hydrocarbons, especially CH_4 and HCl with Si produces 15-20% CH_3SiCl_3 along with $HSiCl_3$ and $SiCl_4$ (at 600-900°C) under pressure.[831]

As mentioned earlier in the context of corrosion of reactor materials, direct synthesis with halogen-free compounds also provides an alternative route of introducing alkyl groups to silicon compound. Thus when dimethylether (or any dialkylether) with dry HCl (molecular ratio 1 : 0.1 to 3) is passed over Si-Cu alloy at 200-500°C, methyl-substituted chlorosilanes are produced.[832] But if lower alcohols are used, tetraalkoxysilanes are obtained (unlike above, here alcohol with silicon do not form ester for further reaction) as follows:

$$4\ CH_3OH + Si \xrightarrow[250^\circ]{Cu} Si(OCH_3)_4 + 2\ H_2$$

Hydrogen thus produced as by-product, further react to form silane (SiH_4), hydrocarbons and water.

Aromatic derivatives of silicon, on the other hand, can be produced by Wurtz reaction:

$$SiCl_4 + 4\ ArCl + 8\ Na = Ar_4Si + 8\ NaCl$$

Another direct synthesis route to alkyl-silicon compounds is by Friedel-Craft's synthesis:

$$SiCl_4 + 2\ Zn(C_2H_5)_2 = 2\ ZnCl_2 + (C_2H_5)Si$$

Diethylzinc, used in above reaction, has also been used in Friedel-Ladenburg reaction to synthesize alkylalkoxysilanes:

$$Zn(C_2H_5)_2 + 2\ Si(OC_2H_5)_3Cl + 2\ Na = 2\ C_2H_5Si(OC_2H_5)_3 + 2\ NaCl + Zn$$

Ladenburg also used diethylmercury (in sealed tube at 300°C) to synthesize phenylchlorosilanes:

$$SiCl_4 + Hg(C_6H_5)_2 = C_6H_5SiCl_3 + C_6H_5HgCl$$

Kippin-Dilthey reaction similarly provides means to directly synthesize organosilicon compounds, by the following type of reactions:

$$SiCl_4 + 2C_2H_5MgCl = (C_2H_5)_2SiCl_2 + 2MgCl_2$$

Organometallic synthesis

In these reactions, the organometallic compound transfers the alkyl-hydrocarbon forming silicon monomeric compounds. The reaction can be represented as:

$$RMe + X\text{–}Si\text{–} = R\text{–}Si\text{–} + MeX$$

Grignard reagent (RMgX) can also be used[833,834] for transferring hydrocarbon groups to silicon atom, according to the following reaction:

$$SiCl_4 + 4RMgX = R_4Si + 4MgClX$$
$$RMgX + XSi\text{–} = RSi\text{–} + MgX$$

In the above reaction, R could be either saturated or unsaturated hydrocarbon:

$$HSiCl_3 + RMgX = MgXCl + RSiHCl_2 + R_2SiHCl + R_2SiH$$
$$HSiCl_3 + ArMgX = MgXCl + HSiArCl_2 + HSiAr_2Cl + HSiAr_3$$

(Ar represents aryl group in above reaction). The Grignard Reagent (RMgX) itself is prepared (aryl or alkylmagnesium-halide) by reaction of an alkyl or aryl halide with magnesium turnings suspended in lower aliphatic ether (e.g. diethylether, dibutylether, THF, ethylene-glycol, etc.). Among these solvents, reactivity is found highest in THF and thus used to introduce vinyl-group using vinylmagnesium bromide to silicon.[835] Grignard reaction is exothermic and often requires cooling. Continuous production process involves filling magnesium-turnings into a stationary bed and irrigating the bed from above with a solution of $SiCl_4$ or silicic-acid ester in a sufficiently high boiling solvent, while passing the organic-halide counter-current fashion from below:

$$SiCl_4 + 4RMgX = R_4Si + 4MgClX$$
$$4RMgX + H_2SiF_6 = R_4Si + 2HF + 4MgXF$$
$$RSi\text{–}X + R'MgX = (RR')Si= + MgX_2$$

Accordingly, trichlorosilane reacts with Grignard Reagent to form triorganosilane:[836]

$$HSiCl_3 + 3RMgX = R_3SiH + 3MgXCl$$

Using this synthesis route, it is possible to synthesize organo-H-halosilanes by partial substitution in trichlorosilane:[837-841]

$$SiHCl_3 + CH_5MgCl = CH_3SiHCl_2 + MgCl_2$$
$$HSiCl_3 + C_6H_5MgBr = C_6H_5SiHCl_2 + MgBrCl$$

Reactivity of Si–H bond with Grignard Reagent, depends on the nature of the solvent;[842] thus while triphenylsilane do not react with phenylmagnesium-bromide in ether, xylene, or a mixture of ether and dioxane, it can be converted into tetraphenylsilane with tetrahydrofuran as solvent (with 14% yield expected). If fluorosilanes are used, above reactions can occur with bulky substitutions as well as the fluorine itself occupies small volume. Silicon bonded alkoxy groups (e.g. di-isobutyldimethoxysilane) can be synthesized from tetramethoxysilane and isobutyl (Grignard Reagent), but the reaction proceeds more slowly than halogens (it gives 95% yield and above 97% purity). From economic point of view, Grignard Reaction is expensive and used only for organosilicon compounds having high sale value. Among the alkoxysilanes, the derivatives of lower aliphatic alcohols, particularly $Si(OCH_3)_4$ and $Si(OC_2H_5)_4$, are most suitable from Grignard synthesis.

Organometallic compounds made from other metals, like aluminium ($RAlCl_2$, R_2AlCl as well as R_3Al) has also been tried successfully for organosilicon synthesis:[843]

$$HSiCl_3 + RAlCl_2 = RHSiCl_2 + AlCl_3$$
$$HSiCl_3 + R_2AlCl = R_2HSiCl + AlCl_3$$

Substitution of organic group to trichlorosilane, makes the reaction proceed more easily and smoothly.

Synthesis with organic compounds of alkali metals

Organic compounds of alkali metals (e.g. RLi), can transfer the organic group to silicon atom. While $SiCl_4$ do not react with Grignard reagents containing tertiary organic groups, t-butylhalosilanes such as t-butyltriclorosilane and di-t-butyldiclorosilane can be synthesized with organolithium compounds:[844]

$$RLi + Cl{-}Si{\equiv} = R{-}Si{\equiv} + LiCl$$
$$SiCl_4 + (\text{t-butyl})Li = (\text{t-butyl})Si$$
$$SiCl_4 + (\text{Cyclohexyl})Li = (\text{tricyclo-hexyl})SiCl$$

Alkoxysilanes[846] and silanes with Si-H bonds can function as reactants in addition to halosilanes. The reaction of trichlorosilane with an organolithium compound in diethylether, therefore, proceeds as far as the tetra-organosilane:[847,848]

$$SiHCl_3 + 4\ Li = R_4Si + LiH + 3\ LiCl$$

With an excess of phenyl-lithium or ethyl-lithium in diethylether, monophenylsilane can be converted into tetraphenylsilane or triethylphenylsilane.[849]

The Si–H bond in $SiHCl_3$ and $SiCl_4$ is more reactive than the Si–Cl bond (needs to be activated by supply of energy during reaction); accordingly following addition/substitution or combination of both may be helpful in certain cases.

Substitution reactions:

$$HSiCl_3 + C_6H_6 = H_2 + C_6H_5SiCl_3$$
$$HSiCl_3 + RCl = HCl + RSiCl_3$$
$$HSiCl_3 + CH_2 = CHCl = CH_2{=}CH\text{–}SiCl_3 + HCl$$
$$HSiCl_3 + CH_2 = CHCH_2Cl = Cl_3Si\text{–}CH_2CH_2CH_2Cl$$
$$SiCl_4 + CH_4 = CH_3SiCl_3 + HCl$$
$$HSiCl_3 + C_6H_5Cl \xrightarrow{500°C} C_6H_5SiCl_3 + HCl$$
$$MeHSiCl_2 + C_6H_5Cl \xrightarrow{500°C} C_6H_5MeSiCl_2 + HCl$$

Above reactions with vinyl chloride ($CH_2 = CHCl$) producing trichlorovinylsilane and similar reactions with allylchloride ($CH_2 = CHCH_2Cl$) are of commercial importance.

Addition reactions:

$$HSiCl_3 + R. = RH + .SiCl_3 \text{ (R. = initiating radical)}$$
$$(CH_2)_3.CH = CH_2 + .SiCl_3 = CH_3(CH_2)_2CHCH_2SiCl_3$$
$$CH_3(CH_2)_2CHCH_2SiCl_3 + HSiCl_3 = CH_3(CH_2)_4SiCl_3 + .SiCl_3$$

Simultaneous to above reactions, self-polymerization reaction of olefins kinetically competes with above addition reactions. Degree of polarization of Si–H bond with substitution of various electronegative group on silicon, plays a decisive role in determining reactivity of Si–H bond. Addition reactions with Si–H bond are of commercial interest, because these reactions yield precisely the theoretically expected product, and the number as well as amount of by-products in relatively small quantities. This is in contrast to direct synthesis, as well as organometallic synthesis route. Only the addition reactions with SiH_4 sometime becomes problematic as it may liberate a number of Si–Si and Si–C polymeric by-products.

Addition reactions of Si–H bond to fluorosubstituted olefins are very suitable for synthesis of fluorosilanes. In such reactions catalysts like platinized-charcoal, chloroplatinic-acid or its salts, tertiary-amines, ozone or even irradiation with light is used to accelerate the synthesis:[850-854]

$HSiCl_3 + (CF_3)_2C{=}CH_2 = (CF_3)_2CHCH_2SiCl_3$
$HSiCl_3 + CF_3CH{=}CH_2 = CF_3CH_2CH_2SiCl_3$
$HSiCl_3 + CFCl{=}CF_2 = HCFCl.CF_2SiCl_3$
$HSiCl_3 + CH_2{=}CHCH_2Cl = ClCH_2CH_2CH_2SiCl_3$

At temperatures above 250°C and in presence of $AlCl_3$, Si–H bond is capable of undergoing condensation reactions forming arylsilanes:

$HSiCl_3 + C_6H_5Cl = C_6H_5SiCl_2 + HCl$
$HSiCl_3 + C_6H_5Cl = ClC_6H_4SiCl_3 + H_2$

$$HSiCl_3 + p\text{–}C_6H_4Cl_2 \xrightarrow[600°C]{\text{in benzene}} p\text{–}ClC_6H_4SiCl_3 + HCl$$

$HSiCl_3 + RC{\equiv}CH = RCH{=}CHSiCl_3$
$HSiCl_3 + CH_2{=}CHSiCl_3 = Cl_3SiCH_2CH_2.SiCl_3$/1,2-bis (trichlorosilyl ethane)

$$HSiCl_3 \xrightarrow[\text{0°C, electric discharge,100 torr}]{\text{vinyltricloro silane}} Si_2Cl_6$$

$$HSiCl_3 + R_2S = O \xrightarrow[\text{250°C, } Et_2O]{} R_2SHOSiCl_3$$

Exchange Reaction (exchange of R with Cl):
Tetraphenylsilane produces phenylchlorosilanes on being heated at 300-400°C (in a sealed tube):[855,856]

$$SiCl_4 + (C_6H_5)_4Si = (C_6H_5)_nSiCl_{4-n} \ (n = 1,2,3)$$

Cu(I) chloride is used as catalyst, and 34 % of $(C_6H_5)_3SiCl$ is formed within 12 h at 300-360°C. $ZrCl_4$ has been used as promoter here.

Exchange of H with R and X:
Substituents on organosilanes having Si–H bond, migrates easily (by disproportionation with heat) resulting in a number of by-products. $AlBr_3$ is used as catalyst for these disproportionation reaction:

$$HSiCl_3 \xrightarrow{\text{disproportionation}} SiH_4 + SiH_2Cl_2$$

$$MeSiCl_3 + Me_3SiCl \xrightarrow{\text{coproportionation}} 2Me_2SiCl_2$$

All the reactions discussed above generates silanes with non-functional organic substituents. We will now discuss in short reactions that generate functional organic substituents, and thereafter producing polymers from these later class of compounds. Silicon with functional organohalo groups is represented by the general formula R_nSiX_{4-n} having at least one Si–X group that can be cleaved hydrolytically.

Hydrosilylation Reactions (addition reactions):

Silylation is the displacement of active hydrogen from an organic molecule by a silyl group. The active hydrogen is usually OH, NH or SH, and the silylating agent is usually trimethylsilylhalide $(CH_3)_3SiCl$ or a nitrogen functional compound. In hydrosilylation reaction an unsaturated hydrocarbon is used and the displaced hydrogen adds up to the unsaturated group:

$$HSiCl_3 + R_2C{=}CR_2 = Cl_3SiCR_2CR_2H$$
$$HSiCl_3 + RC{\equiv}CR = Cl_3SiCR{=}CRH$$

Silyl radicals can be generated by heat, by decomposition of radical initiators (e.g. acyl-peroxides, azonitriles), by UV-light, or by gamma-radiation. Transition-metals are the best catalyst for hydrosilylation reaction. Platinum on active charocoal is an effective catalyst for the reaction[857] of $HSiCl_3$ with acetylene, ethylene, butadiene, allyl chloride and $H_2C = CF_2$ at about 130°C. Chloroplatinic-acid and its salts are reported to bring this catalytic reaction down to room temperature and below,[858] which provides the opportunity to hydrosilylation in an economic way. Hydrosilylation is used in industry for synthesizing alkylsilanes and functional silanes, cross-linking silicon polymers, binding silicon polymers to organic polymers, and removing impurities that contain Si–H group from silane fractions of the direct synthesis route, mentioned earlier.

Organopolysilanes in which organosilyl groups are linked by Si–Si bond, are used as starting material for silicon carbide ceramics, as photoresists in microelectronics, as photoinitiators, as non-linear optical devices. Disilanes are also being used to develop industrially important Si–C compounds and *in situ* cleavage of hexamethyldisilane with solution of elemental iodine producing light sensitive iodotrimethylsilane.

Alkoxy- and Aryloxy-silanes

These silanes with reactive alkoxy- or aryl-oxy groups finds good industrial use as they do not produce any HCl, which on the one hand avoids corrosion of reaction vessel and on the other allows use of acid sensitive substances like acetone or basic catalysts in the reaction medium. They also possess higher boiling point and flash point than corresponding chloro compounds, which makes possible use of open apparatus for their use. Further they hydrolyze slowly, which makes their control easy and thus can be easily converted to 'silanols', silicon resins or sol-gel precursors.[859] Organo (organoxy) silanes are prepared by reaction of organohalosilanes with alcohols or alkoxides:

$$\equiv Si{-}X + ROH = \equiv Si{-}OR + HX$$
$$\equiv Si{-}X + ROM = \equiv Si{-}OR + MX$$

Pyridine or tertiary-amines may be added to the reaction to remove acid formed. Very high yield of alkoxysilanes are obtained by reacting[860] chlorosilanes with alkylorthoformates in the presence of $AlCl_3$. The Si–H bond can also be converted to Si–OR bond by catalyzed reactions with alcohols or alkoxides, and with aldehydes, ketones and ethers. In these reactions, the best arrangement involves counter-current gas-liquid reaction in a corrosion resistant distillation column, where mixtures of chlorosilanes and alcohol is fed from top of the column, while vapour of alcohols is pushed up the column from the bottom.[861] This solvent-free continuous production is the most attractive industrial route.

Similar group of compounds having Si–O link (although in the strictest sense not organosilanes) are silicon-ethers and –esters (≡SiOR). They are the esters of orthosilicic-acid $Si(OR)_4$ and lower derivatives. Unlike organo (organoxy) silanes, tetraalkoxysilanes are prepared by direct reaction of silicon,[862] or suitable natural silicates[863] and alcohols. These reactions produce a large number of by-products, and thus are used where gross physical property of the product is used as the final material for economic advantage. Thus the product of ($SiCl_4 + C_2H_5OH$ + 40 % silica in water) finds a number of applications like protection of metallic component of bridges, moulds for precision casting, renovation of weathered sandstones, etc.

Application of alkoxysilanes depends on whether the Si–OR bond will remain intact or will be hydrolyzed on application. Hydrolysis applications are in binder for mould sands in investment and thin shell casting, binder for refractories, resins, coatings, low heat glass, cross-linking agents, and as adhesion promoter. Applications where Si–OR bond remains intact are lubricants, heat transfer and hydraulic fluid, dielectric medium, fluid for diffusion pump, etc. The major difference between alkoxysilane and silicones is the susceptibility of Si–OR bond to hydrolysis. In accordance with the more pronounced covalent character of the bond compared with Si–X bond, sensitivity of Si–OR bond to hydrolysis is considerably less. Ease of hydrolysis also depends on the organic substituent (R) in the molecule, as

$$Si(OC_2H_5)_4 + 2\,H_2O \xrightarrow{\text{acid or base}} SiO_2 + 4\,C_2H_5OH$$

The esterification reaction for its preparation has been described by Von-Ebelman in 1846, as follows:

$$SiCl_4 + 4\,C_2H_5OH = Si(OC_2H_5)_4 + 4\,HCl$$

In general, $R_{4-n} + n\mathrm{ROH} = R_{4-n}Si(OR)_n + n\mathrm{HCl}$

Tetraethoxysilane and its polymeric derivative accounts for more than 90 % of the world sale value of non-aryl or alkyl-substituted esters. In 1980, its cost was $2.40-3.50/kg, and price of nonethyl-esters were $4.50-100/kg.

The acyloxysilanes are produced by reaction of an anhydride and a chlorosilane[864] as:

$$SiCl_4 + R(R.C = O)_2O = 4\,Si(OCOR)_4 + 4\,RCOCl$$

An analogous reaction involving anhydrides and alkoxysilanes also produces acyloxysilanes. The reaction of acids directly with chlorosilanes is usually not carried out in practice. Marked improvement in the yield of this reaction has been reported with addition of a small amount of acetic-anhydride or EDTH.[865] Methyltriacetoxysilane is commercially the most important acyloxysilane. In the USA, bulk of the acyloxysilanes produced is used captively for silicon rubber production by major producers like General Electric and Dow-Corning. Union Carbide and Petrarch Systems also manufactures acyloxysilanes. Some organoalkoxysilanes are prepared by reacting corresponding silanes with polyhydric alcohols (e.g. ethylene-glycol, glycerol, etc.[866-867] These products are soluble in water:

$$(CH_3)_2Si(OCH_2)_2\text{–}CHOH \text{ (glycerol product)}$$

Acyloxysilanes, $R_nSi(OCOR')_{4-n}$ are less readily formed than alkoxy compounds and they are important industrial materials used as cross-linking agent in one pack of silicon-sealants (RTV) system. They are produced commercially by reaction of chlorosilanes with acetic-anhydride:

$$RSiCl_3 + 3\ Ac_2O = RSi(OAc)_3 + 3\ AcCl$$

These compounds known as acetic acid hardeners, are readily hydrolyzable. Oximes also are easily hydrolysable and commercially important materials.

Among ether-functional monomers, those containing epoxy group (e.g. $CH_2 \overset{O}{\diagdown\diagup} CH\text{–}$) are of commercial interest as silanes and siloxanes with epoxy-group as substituent are used for preparation of modified siloxanes.

Nitrogen compounds

A number of industrially important nitrogenous silicon compounds, like aminosilanes, amidosilanes, silazanes, cyanoorganosilanes, and silatranes, have been synthesized, basic principle for synthesis of which we will discuss here. The basicity of amino group in aminoalkylsilane, increases with substituents like trimethylsilyl group, etc., while it decreases by phenyl or siloxane oxygen. These compounds are prepared by reaction between haloorganosilane and ammonia or amines:

$$\equiv SiCl + NH_3 = \underset{\text{silylamine}}{\equiv SiNH_2} + HCl$$

A 20-fold excess ammonia needs to be added to obtain primary amine. Addition of sodamide in liquid ammonia breaks the Si–C bond and produces a silicon functional amino-compound instead of an organofunctional one,[868] as:

$$ClCH_2(CH_3)_3Si + NaNH_2 = (CH_3)_3SiNHCH_3 + NaCl$$

Amidosilanes have a structure, $R_nSi\ NRC(O)R_{4-n}$ and are used as neutral cross-linking agent. These compounds are formed by reaction between chlorosilanes and organic amides,[869] with a strong HCl acceptor in the solution (e.g. sodium metal or methoxide). The silicon-monofunctional derivatives of urea and acetamide are used in large quantities for pharmaceuticals.

Silazanes having the general structure ≡Si–NH (mono), $(\equiv Si)_2NH$ (di), and ≡Si–NH–Si≡ (poly), are produced by reaction between chlorosilanes and ammonia. They produce polymers having repeatitive units of SiNHSi. Cyano-organosilanes (organofunctional silanes with CN group) are produced by addition of Si–H bond to unsaturated nitriles, generally acrylo nitrile:

$$\equiv SiH + CH_2{=}CH{-}CN = \equiv SiCH_2CH_2CN \text{ or } \equiv SiCH(CH_3)CN$$
$$SiHCl_3 + CH_2{=}CH{-}CN = NC.CH_2.CH_2.SiCl_3$$

Various amides are used as catalyst in later reactions and a yield of 70-75 % is achieved. These compounds are generally used as intermediates for other reactions (e.g. preparation of carboxy-functional silanes). Beta-cyanoethyl and gama-cyanopropylmethylsiloxane units inserted into dimethylsiloxane chain imparts swelling resistance property to the rubber. These are also used to increase dielectric constant of silicon oils (in electrical applications).

Silicon-Metal bond

Organosilicon compounds having Si–M bond are very reactive towards oxygen, CO_2 and air; thus they are prepared under nitrogen with appropriate solvent. Below are some of their preparative method:

Simplest compound silicides: $Si + M \xrightarrow{600\text{-}700°C} M_2Si$ (M = Li, Na, K, Rb, Cs)

Breaking Si–H bond: $(C_6H_5)_3SiH \xrightarrow{Na/K} (C_6H_5)_3SiK$

Breaking Si–C bond: $(C_6H_5)_3SiC(C_6H_5)_3 + 2K = (C_6H_5)_3SiK + (C_6H_5)C_3K$

Breaking Si–Si bond: $(C_6H_5)_3Si.Si(CH_3)_3 \xrightarrow{Na/K} (C_6H_5)_3SiK + (CH_3)_3SiK$

Breaking Si–O–C and Si–O–Si bonds: $(C_6H_5)_2Si.O.C_2H_5 \xrightarrow{Na/K} (C_6H_5)SiK + C_2H_5OK(C_6H_5)_3Si{-}O{-}Si(C_6H_5)_3 + 2Li = (C_6H_5)_3SiLi + (C_6H_5)_3SiOLi$

Breaking Si-halogen bond: $(C_6H_5)_3SiCl \xrightarrow{Na/K} (C_6H_5)_3SiK + KCl$

Note that in all the above reactions for formation of Si-M bond, the silicon compound must contain the phenyl group associatd with it, otherwise Si-M bond formation is not possible.

Preparation of polymeric silicon compounds

Commercially important polymeric silicon compounds are: polyorganosiloxanes and silicon fluids.

Polyorganosiloxanes:

Both linear and cyclic polyorganosiloxanes can be prepared from monomeric silicon functional organosilanes, preparations of which has been described earlier. Among all the processes available for such polymerization reaction, hydrolysis route is the simplest as well as an economical route, and accordingly all commercial practice follow this method only. Hydrolysis first produces silanols, which because of the instability of silanol bond tend to polymerize to polysiloxanes:

$$RSiX_3 \rightarrow RSi(OH)_3 \rightarrow RSiO_{3/2} \text{ (siloxane unit)}$$
$$SiX_4 \rightarrow Si(OH)_4 \rightarrow SiO_{4/2} \text{ (siloxane unit)}$$

Mixtures of oligomeric siloxanes thus produced can be converted either entirely to cyclic siloxanes or to polymeric linear-polysiloxanes. Most of the industrial productions start with methylclorosilanes. Rate of hydrolysis increases with polarity of Si-X bond and the number of X atom attached to each silicon in the monomer. For larger bulky molecules, steric hindrance also determines the ease of reaction. Since these reactions are basically exothermic in nature (due to dissolution of HCl in water), the mixture during course of the reaction needs to be constantly cooled:

$$2(CH_3)_3SiCl + H_2O = (CH_3)_3Si\text{–}O\text{–}Si(CH_3)_3 + 2HCl$$

Cyclization tendency has been found to increase as the solution is made more acidic, while excess water produces highly cross-linked gel like mass or pulverulent polymer. If water-soluble organic solvents like THF or dioxane is used, the yield is mostly low molecular weight polysiloxanes. On the other hand, adding immiscible or partly miscible organic solvents like toluene, xylene, diethylether, dibutylether, and tricloroethylene, results in taking up of the product as they form by hydrolysis into the organic phase, thereby protecting the product from the action of resultant acid and simultaneously separating the product from aqueous phase. Hydrolysis can also be carried out with ether, or in gas phase. Ratio of linear to cyclic compound in final product, can be controlled by choosing the variable hydrolysis conditions. Thus quick removal of HCl as soon as they are formed, results in more of short-chain siloxanediols, whereas prolonged exposure to HCl produces more of cyclic-siloxanes. If the condition of hydrolysis is maintained alkaline, mostly higher polymers are obtained, whereas addition of diethylether produces mainly lower cyclic polymers.

The other commercially important hydrolysis method to produce polysiloxans is methanolysis. Thus reactions of halosilanes with methanol produces polysiloxanes and methylchlorode, as:

$$(CH_3)_2SiCl_2 + CH_3OH = HO\ (CH_3)_2SiO_n\ H + CH_3Cl + H_2O$$

In these processes, recycling of chlorine as cloromethane avoids HCl waste.

Mixture of condensed and monomeric product produced by above hydrolysis step, further needs to be chemically treated, in order to make the cyclization and polymerization step complete. Pure cyclic siloxanes are thus produced by completing cyclization reaction. Cyclo compounds like octamethylcyclotetrasiloxane and decamethylcyclopentasiloxanes are major industrial products which are sold as such, or used to produce polydimethylsiloxanes. Cyclization reaction is completed by heating the product mix from hydrolysis step with KOH. Pure linear polyorganosiloxanes can be prepared also from mixture of its cyclic products by anionic or cationic catalysts made with alkali metals. On the other hand, mixtures of linear oligomers derived from methanolysis step, can be converted to pure linear polymers by acid catalysts such as polyclorophosphazenes.[870-872]

As mentioned earlier, linear polydimethylsiloxane fluids are of great commercial importance and in some respect superior to silicon fluids. In these polydimethylsiloxanes, the end group determines their use; for example, trimethylsilyl terminated PDMS (polydimethylsiloxane) are typical silicon-fluids, whereas hydroxy- and vinyl-terminated polymers find major applications in silicon rubbers. Siloxane based copolymers (as described earlier) have great commercial value. Physical state of PDMS also depends on its degree of polymerization; for example, di-polymers are in liquid state until about 30 % of tri-units are in the compound (CH_3/Si about 1.7). A region of resinous polymer appear with more than about 50 % of tri-units and extends up to 90 % tri-unit (10 % di-unit). Beyond this ratio the product becomes slowly glassy and brittle (gel-like character), especially CH_3/Si ratio falls below 1. Polyethylsiloxanes on the other hand, are softer than polymethylsiloxanes with the same degree of cross-linking, and here resinous character develops with 50% di- and 50 % tri-units (i.e C_2H_5/Si ratio of 0.5).[873-874] Heat resistance property of the polymers simultaneously decreases rapidly with increasing size of alkyl group in it. Incorporation of phenyl group in the structure not only improves mechanical and electrical properties of the polymers, but long-term heat stability of the product. Bond energies, which determines stability of these compounds, are presented[875] in Table 10.

Silicones

These compounds have a repeating Si–O chain, with a significant portion

Table 10: Bond energy of some silicon polymer bond[875].

Bond	Bond Energy
Si–Si	42.2 kcal/mole
Si–C	69.3 kcal/mole
Si–H	70.4 kcal/mole
Si–Cl	85.7 kcal/mole
Si–O	88.2 kcal/mole
Si–F	129.3 kcal/mole

of organic group (R) attached to the silicon atom. They are represented by the general formula:

$$[(R_nSiO_{4-n/2}]_m, \text{ where } n = 1\text{-}3 \text{ and } m \geq 2$$

Thus these structures can be distinguished from other silicon polymers whose chain links are shown below:

≡Si – Si≡	≡Si–O–Si≡	≡Si–NH–Si≡	≡Si–S–Si	≡Si–$(CH_2)_n$–Si≡
Polysilane	Polysiloxane	Polysilazane	Polysilthiane	Polysilakylene

≡Si–C_6H_4–Si≡
Polysilarylene

≡Si–$(CH_2)_n$–Si–O–Si≡
Polysilakylenosiloxane

≡Si–C_6H_4–Si–O–Si≡
Polyarylenosiloxane

≡Si–$(CH_2)_n$ –Si–Si≡
Polysilakylenosilone

Commercial silicones have mostly R as methyl group, and other groups like larger alkyls, fluoroalkyls, phenyls, vinyls and a few other groups are attached for specific purposes only. Special properties of silicon-fluids like good thermal stability, good dielectric property, lubricating properties, low temperatuure dependence of most of its properties, strong hydrophobicity, release property, etc. makes them of great commercial value. Besides these, research has also focused on silicon polymers with heteroatoms (other than oxygen), but physicochemical properties of these heterosilicon-compounds has never been found, in general, superior to silicones. Silicon-organic copolymers generated from organofunctional silanes have also found commercial interest for their good physical strength, durability, and low cost of production. These later class of compounds are generally prepared by reaction of siloxanes containing silanol or alkoxy functions with the hydroxy groups of organic polymers, forming Si–O–C link chain in the resultant polymer (e.g. silicon alkyds). These reactions can also be extended to polyesters, epoxy, phenol-formaldehyde, and acrylic compositions. Theoretically, although such Si–O–C bond should be prone to easy hydrolysis, in practice such polymer shows good stability. Other hydrocarbons that can be attached to siloxane group are polyethers (forming

$\equiv SiCH_2O(CH_2CH_2O)_nR$ polymer,[876] polycarbonates by reaction with phosgene (forming $\equiv SiO(C_6H_4)C(CH_3)_2(C_6H_4O).CO$, and polyimides. Besides these many other siloxane-organic copolymers have been synthesized. Dimethyl silicone fluids are prepared by catalyzed equilibration of the crude raw material with a source of the chain terminator. $(CH_3)_3Si_{0.5}$, which produces MD_nM and D_m polymers.[877]

Some additional direct synthesis reactions with $HSiCl_3$ and $SiCl_4$

1. $SiCl_4 + (C_6H_5)_3SiLi = (C_6H_5)_3Si.Si(C_6H_5)_3 + (C_6H_5)_3SiH$ (yield = 72.5%)
2. $SiCl_4 + (C_6H_5)_3SiK = (C_6H_5)_3SiSiCl_3$ (yield = 27%)
3. $HSiCl_3 + (C_6H_5)_3SiLi = (Ph_3Si)_3SiH + Ph_3SiSiPh_3 + Ph_3SiH$ (yield = 4.4%)
4. $SiCl_4 + Me_3SiCH_2Li = (Me_2SiCH_2)_2SiCl_2 + (Me_2SiCH_2)_2SiCl$ (yield = 50%)
5. $SiCl_4 + Me_3SiCH_2Li = (Me_2SiCH_2)_3SiCl$ (yield = 32%)
6. $SiCl_4 + Me_3SiCH_2Li = (Me_2SiCH_2)_4Si$ (yield = 33%)
7. $HSiCl_3 + Me_3SiCH_2Li = (Me_2SiCH_2)_3SiH$ (yield = 71%)
8. $SiCl_4 + (Me_3Si)_2NLi = Cl_3SiN(SiMe_3)_2$
9. $SiCl_4 + (Me_3Si)_2NK = Cl_2SiN(SiMe_3)_2$
10. $SiCl_4 + [(CH_3O)_3Si]_2N.Na = Cl_3SiN[Si(OCH_3)_3]_2$
11. $SiCl_4 + [(C_2H_5O)_3Si]_2N.Na = Cl_3SiN[Si(OC_2H_5)_3]_2$
12. $SiCl_4 + [(iso–C_3H_7O)_3Si]_2N.Na = Cl_3SiN[Si(OC_3H_7–iso)_3]_2$
13. $SiCl_4 + (Me_2CH_2=CHSi)_2N.Na = Cl_3SiN(SiCH=CH_2Me_2)_2$
14. $SiCl_4 + Me_2Si(NLiCH_3)_2 = MeSi\langle(NMe)(NMe)\rangle Si\langle(NMe)(NMe)\rangle SiMe_2$
15. $SiCl_4 + (CH_3)_2Si(NLiC_6H_4CF_3–m)_2 = Si\langle(NC_6H_4CF_3–m)(NC_6H_4CF_3–m)\rangle SiCl_2$
16. $SiCl_4 + (CH_3)_2Si(NLiC_6H_5)_2 = Me_2Si\langle(NC_6H_5)(NC_6H_5)\rangle SiCl_2$
17. $SiCl_4 + CH_3N(CH_3)_2SiNLi(CH_3)_3 = CH_3N\langle(Si(CH_3)_2N(CH_3))(Si(CH_3)_2N(CH_3))\rangle SiCl_2$
18. $SiCl_4 + (Me_3SiNLi(CH_2)_2 = (CH_3)_3SiN$–$(CH_2)_2$–$N.Si(CH_3)_3$, both N bonded to a central Si, which is also bonded to $(CH_3)_3SiN$–$(CH_2)_2$–$NSi(CH_3)_3$

19. $SiCl_4 + (CH_3)_3SiONa = Si(OSiMe_3)_4$
20. $HSiCl_3 + (C_6H_5)_2SiONa = (C_6H_5)_3SiOSiHCl_2$
21. $HSiCl_3 + (C_6H_5)_2SiONa = (Ph_3SiO)_3SiH$
22. $SiCl_4 + (C_6H_5)_2SiONa = (C_6H_5)_3SiOSiCl_3$
23. $SiCl_4 + (C_6H_5)_2SiONa = (C_6H_5)_3SiOSiCl_3$
24. $SiCl_4 + (C_6H_5)_2SiONa = (Ph_3SiO)_4Si$
25. $SiCl_4 + NaO(C_2H_5Ph.SiO)_nNa$ = Polymer
26. $SiCl_4 + (CH_3)_3SiCH_2MgCl = (Me_3SiCH_2)_3SiCl + (Me_2SiCH_2)_4Si$
27. $SiCl_4 + (CH_3)_3SiCH_2MgCL = (CH_3)_3SiCH_2SiCl_3$
28. $HSiCl_3 + Me_3SiCH_2MgBr = Me_3SICH_2SiHCl_2 + (Me_3SiCH_2)_2SiHCl + (Me_3SiCH_2)_3SiH + (Me_3SiCH_2)_3SiH.$
29. $HSiCl_3 + 4RLi = R_4Si + LiH + 3LiCl$
30. $RSiCl_3 + RAlCl_2 = R_2SiCl_2 + AlCl_3$
31. $RSiCl_3 + R_2AlCl = R_3SiCl + AlCl_3$
32. $HSiCl_3$ + But-1-ene = n-Butyltrichlorosilane.
33. $HSiCl_3$ + But-2-ene = s-Butyltrichlorosilane.
34. $HSiCl_3$ + n-Hexene = n-Hexyltrichlorosilane.
35. $HSiCl_3$ + n-Hexadecene = n-Hexadecyltrichlorosilane.
36. $HSiCl_3$ + n-Octadecene = n-Octadecyltrichlorosilane.
37. $HSiCl_3$ + Hexa-1, 5-diene = Bis-trichlorosilyl-hexane.
38. $HSiCl_3 + C_6H_6 \xrightarrow[\text{catalyst}]{BCl_3} C_6H_5SiCl_3 + HCl$
39. $HSiCl_3 + C_6H_5Cl = C_6H_5SiCl_3 + HCl$
40. $SiCl_4 + CH_4 \xrightarrow{\text{powder Si}} CH_3SiCl_3 + HCl$ (bound by powder-Si)
41. $HSiCl_3 + RMgCl = RSiHCl_2 + MgCl_2$
42. $(C_6H_5)_4Si + AlCl_3 = (C_6H_5)_3SiCl + C_6H_5AlCl_2$
43. $(C_6H_5)_3SiCl + AlCl_3 = (C_6H_5)_2SiCl_2 + (C_6H_5)AlCl_2$
44. $HSiCl_3 + CH_2{=}CHCH{=}CH_2 \xrightarrow{Pt} CH_3CH{=}CH\ CH_2SiCl_3 \xrightarrow[\text{excess}]{HSiCl_3} CH_3CH(SiCl_3)(CH_2)_2SiCl_3$
45. $HSiCl_3 + RC{\equiv}CH \xrightarrow{Pt/C} RCH{=}CHSiCl_3$ (*trans*-product)

    $\xrightarrow{\text{Peroxide}} RCH{=}CHSiCl_3$ (*cis*-product)
46. $HSiCl_3 + CH_2{=}CH{\cdot}COOCH_2CH{=}CH_2 \rightarrow CH_2{=}CH{\cdot}COO(CH_2)_3{\cdot}SiCl_3$
47. $HSiCl_3 + CH_2{=}CHCl{=}CH_2{=}CH{\cdot}SiCl_3 + HCl$
48. $HSiCl_3 + CH_2{=}CH{\cdot}SiCl_3 = Cl_3Si{\cdot}(CH_2)_2{\cdot}SiCl_3$
49. $HSiCl_3 + CH{\equiv}C{\cdot}CH{\cdot}CH_2 = CH_2CHCH{=}CH{\cdot}SiCl_3$
50. $HSiCl_3 + CH_2{=}CH_2 = CH_2{=}CH{\cdot}SiCl_3 + H_2$
51. $HSiCl_3 + (CF_3)_2C{=}CH_2 = (CF_3)_2CHCH_2SiCl_3$

52. $(C_2H_5)_4Si + CCl_4 \xrightarrow{\text{UV light}} CH_3CHCl(C_2H_5)_3Si + CHCl_3$
53. $HSiCl_3 + CH_2{=}CHCN = NCCH_2CH_2SiCl_3$ (yield = 75%)
54. $HSiCl_3 + R_2S{=}O = R_2S + HOSiCl_3$
55. $HSiCl_3 + n\text{-}Pr_3N = n\text{-}Pr_3NH^+SiCl_3^-$
56. $HSiCl_3 + n\text{-}Pr_3N + R_2CO = R_2CHSiCl_3 + n\text{-}Pr_3NHCl + SiCl_2O$ (R = aryl)
57. $HSiCl_3 + n\text{-}Pr_3N + RCOCL = RCH(SiCl_3)_2 + n\text{-}Pr_3NHCl + SiCl_2O$ (R = alkyl, aryl)
58. $HSiCl_3 + n\text{-}Pr_3N + RX = RSiCl_3 + n\text{-}Pr_3NHCl$ (R=alkyl, aryl and X = Cl,Br)
59. $HSiCl_3 + n\text{-}Pr_3N + RCOOH = RCH_2SiCl_3 + n\text{-}Pr_3NHCl + SiCl_2O$ (R=aryl)
60. $SiCl_4 + (Me_2N)_3P{=}O = SiCl_4{\cdot}2(Me_2N)_3P{=}O$
61. $SiCl_4 + Et_4NCl = (Et_4N)SiCl_5$
62. $SiCl_4 + PhCOCH_2COOEt = (Si(PhCOCH_2COOEt)_3)\ HCl_2$
63. $HSiCl_3 + Fe(CO)_5 = cis\text{-}Cl_3SiFe(H)(CO)_4$
64. $HSiCl_3 + PhC{\equiv}CH \xrightarrow[150\ °C]{\text{MeCN at}} Cl_3SiCPh{=}CH_2 + PhHC{=}CHSiCl_3$
65. $HSiCl_3$ + (2-chlorobiphenyl) = (2-trichlorosilylbiphenyl) + (biphenyl) + $SiCl_4$
66. HSiCl + (1,3-dimethylcyclohexadiene, Me, Me) $\xrightarrow{670°C}$ (3,5-dimethylphenyl)–$SiCl_3$
67. $HSiCl_3$ + (2-methylnaphthalene, Me) $\xrightarrow{670°C}$ (2-naphthyl)–$SiCl_3$ + (2-naphthyl)–CH_2SiCl_3
68. $HSiCl_3$ + (1-methylnaphthalene, Me) $\xrightarrow{670°C}$ (naphthalene bridged by Cl_2Si-CH_2) + (1-naphthyl)–$SiCl_3$
69. $HSiCl_3 + MeOCl + Pr_3N \xrightarrow[\text{reflux}]{\text{MeCN}} MeCH(SiCl_3)_2$ (yield = 55%)
70. $2HSiCl_3 + PhCOPh = Ph_2CHSiCl_3$ (yield = 95%)
71. $HSiCl_3 + AR{=}S.O = \begin{matrix} S...O^2 \\ \vdots \quad \vdots \\ H...SiCl_2 \\ Cl \end{matrix} = {=}SH^+ + {}^-OSiCl_3$

72. $HSiCl_3$ + [10-(3-dimethylaminopropyl)phenothiaphosphine 10-oxide: S, P(=O)$(CH_2)_3NMe_2$] = [10-(3-dimethylaminopropyl)phenothiaphosphine: S, P–$(CH_2)_3NMe_2$]

73. $HSiCl_3 + RCOOR' \xrightarrow{\text{gamma irradiation}} RCH_2OR'$

74. $HSiCl_3 + ArCOOH = ArCH_2SiCl_3 \xrightarrow{\text{KOH.MeOH}} Ar\cdot Me$

75. $(C_6H_5)_3SiH + 2K = (C_6H_5)_3SiK + KH$

76. $(C_6H_5)_2SiH_2 + K = C_6H_5SiH_2K + C_6H_5K$

77. $(C_6H_5)_3SiCl + 2K = (C_6H_5)_3SiK + KCl$

78. $(C_6H_5)_3SiOCH_3 + 2K = (C_6H_5)_3SiK + KOCH_3$

79. $(C_6H_5)_3SiH + 2K = (C_6H_5)_3SiK + KH$

80. $(C_6H_5)Si(CH_3)_2(C_6H_5) + 2K = (C_6H_5)_3SiK + (C_6H_5)C(CH_3)_2K$

81. $(C_6H_5)_3Si\cdot Si(C_6H_5)_3 + 2K = 2(C_6H_5)_3SiK$

82. $(C_6H_5)_2(CH_3)SiK + C_6H_5Br = C_6H_5\cdot CH_2Si(C_6H_5)_2(CH_3)$ (yield = 90%)

83. Ph_2MeSiK + 2 cyclohexanone $\xrightarrow{\text{CuCN}}$ [3-(SiPhMe)cyclohexanone] (yield = 70%)

84. Ph_2MeSiK + Ethylcinnamate $\xrightarrow{\text{CuCN}}$ [$MePh_2Si$–CH(Ph)–CH_2–C(=O)OEt] (yield = 74%)

85. Ph_2MeSiK + Benzoylchlorode $\xrightarrow{MnI_2}$ [cyclohexyl–C(=O)–SiPhMe] (yield = 74%)

86. Ph_2MeSiK + Cyclohexanonecarboxylic acid chloride $\xrightarrow{VCL_3}$ [cyclohexyl–C(=O)–] $SiPh_2Me$ (80 %)

87. $(C_6H_5)_3SiK + C_6H_5Br = (C_6H_5)_4Si + KBr$

88. $(C_6H_5)_3SiK + HCl = (C_6H_5)_3SiH + KCl$

89. $(C_6H_5)_3SiK + (C_6H_5)_3CH = (C_6H_5)_3SiH + (C_6H_5)_3CK$

90. $(C_6H_5)_3SiK + HCHO + H_2O = (C_6H_5)_3SiCH_2OH + KOH$

91. $(C_6H_5)_3SiK + (C_6H_5)_3SiOH = (C_6H_5)_4Si$ + other products

92. $(C_6H_5)_3SiK + (C_6H_5)_3SiH = (C_6H_5)_4Si$ + other products

93. $(C_6H_5)_3SiK + (CH_3)_3SiCl = (C_6H_5)_3SiSi(CH_3)_3 + KCl$

94. $(C_6H_5)_3SiLi + ArOCH_3 \xrightarrow{H^+} (C_6H_5)_3SiCH_3 + ArOH + Li$

95. $(C_6H_5)_3SiLi + C_6H_5CO\cdot H = (C_6H_5)_3SiOCH(Li)C_6H_5 \xrightarrow{H_2O} (C_6H_5)SiOCH_2C_6H_5$

CHAPTER 6

Future Production and Marketing Trend (Alliance Strategy)

With respect to semiconductor/photovoltaic grade silicon. \$2.0/peak watt for silicon module is considered the threshold price for silicon photocells to enter a large domestic market, which still has not been exploited[878]. This cost at present being around \$4/peak watt, a lot of exercise still needs to be done to bring silicon photocell cost to above threshold figure. Reaching this threshold figure will really bring about a sea change in power sector through decentralized power in domestic scenario and environment quality[884]. Simultaneously this price is being closely contested by non-silicon photocells. But from durability point of view and in electronic sector, still there is no replacement for crystalline silicon. Automation (like printing cell) through thin film photocell probably will be able to reach above threshold target in near future.

In the area of silicon compound, exact world consumption of these products is still difficult to obtain. But it is estimated that about 50 % of the world production[879] comes from the USA, followed by Europe at 30 % and Japan at 15%. The production in the USA also increased exponentially from a mere 1000 ton/annum in 1950 to 14000 ton/annum in 1965, to 2×10^5 ton/annum in 1980. Current total silicone production worldwide is estimated at around 4×10^5 ton/annum. These silicones are mainly in the form of fluid, resin, or elastomers. Average cost of fluids in the USA is around \$5-6/kg, that for resins is around \$6-8/kg and elastomers almost at the same level of \$6-8/kg. These main products again are available in 1000 different grades. Monomeric organosilicons are now generally made through direct synthesis and total output is to the tune of 10^6 ton/annum in 1991. As much as 95% of these monomers are utilized for silicone production. To make it cost competitive, in the USA, the monomers are captively consumed by the most large producers. This production technique is followed by Dow-Corning, General Electric and Union-Carbide in the USA, Bayer and Huls, and Wecker Chemetronics in Germany, Rhone-Poulenc in France, Shin-Etsu, Torray and Toshiba in Japan. Price of high value silane in 1991 was around 5-30 $/kg and special silanes (e.g. coupling-agents, cross-linking agents, surface treatment fluids, etc.) costs more. Companies in Europe produces mostly these later variety of chemicals. Other products, of minor importance, produced from monomeric organosilanes are silylating agents, which are mostly consumed by pharmaceutical companies.

Among organosilanes, methyldiclorosilane is of great commercial interest, while among alkyl-substituted (non-aryl) esters, tetraethoxysilane is of greater commercial value. Nonethyl-esters cost around $4.50-100/kg. The largest producer of alkyltrialkoxysilane is Union Carbide and, the next largest Dow-Corning. The bulk of acyloxysilane is produced and used captively by silicon-rubber manufacturers in the USA. On the other hand, the largest consumer of polymethylhydrosiloxane (hydrolysis product of methyldiclorosilane) is the textile industry and next leather as well as silane/silicon producers. Current price of methyldiclorosilane is about $ 4/kg and that of its hydrolysis product, polymethylhydrosiloxane, about $ 7/kg. Cost of other speciality silicon-hydrides (silanes) like dimethylclorosilane, tetramethyldisiloxane and various silanes containing silicon copolymers are around $30-200/kg. A 1 kg dimethyldichlorosilane yields about 0.5 kg of dimethylsiloxane.

As there were only a few handful silicon-compound manufacturers in 1960s, product costs were relatively higher. But subsequently, especially in the last 10 years or so, with the emergence of a number of silicon manufacturers worldwide, especially in Asian countries, cost of silicon products have come down considerably and facing a stiff competition in the market. This trend is likely to continue for still some more time to come. The reason behind this being that cost of raw material as well as finished material production cost in developing countries is much less than that in advanced countries. Because of this reason cost of down-stream silicon compounds manufactured from the above plant intermediates are also cost effective. In fact one of the main reasons for the decline in global silicon based end products in the last one decade has been due to the above reason. This in turn has generated stiff competition in the world market for these materials. One of the fallouts of above competition in price has been strategic alliances among various manufacturers. These alliances has particularly evolved in recent times due to several strategic influences and drives including strong demand growth in the market, globalization of industries, decreasing price of silicon-based end products and stricter quality requirements. The main objective of these alliances is to provide a win-win framework between alliance partners, which inturn will ensure a long-term business for both the associated companies. Dr. B. May James at Dow-

Corning, in his deliberation[880] at the last Silicon Chem Ind IV Conf (held in 1998), stressed upon this point only. This situation in production and marketing is envisaged to continue for another decade.

Simultaneously, research work has also been initiated to radically change the starting raw material for all silicon-compound synthesis, but at a competitive cost than that being pursued at present. One such attempts recently disclosed, is the process pursued by Ritscher[881] at WITCO Corporation (Organosilicones Group), at Friendly, West Virginia (USA). His suggestion has been to switch over to trimethoxysilane from trichlorosilane. Production of this basic raw material has been from a direct reaction between methanol and silicon. Commercially methanol is available at a cheap rate and moreover, the process avoids use of halogen compound and avoids consequent corrosion problem. Above worker has set up a bench scale to commercial size unit at his research facility, and worked out techno-economic feasibility of the whole process. With these basic raw materials, he extended his work to downstream hydrosylilation facility to derive silicon-based end products. The decade long research in this new idea finally culminated in setting up a full-scale commercial unit in 1997. However, other direct synthesis routes suggested earlier, based on halogen-free compounds (e.g. silicon and dimethyl-ether) to produce universal

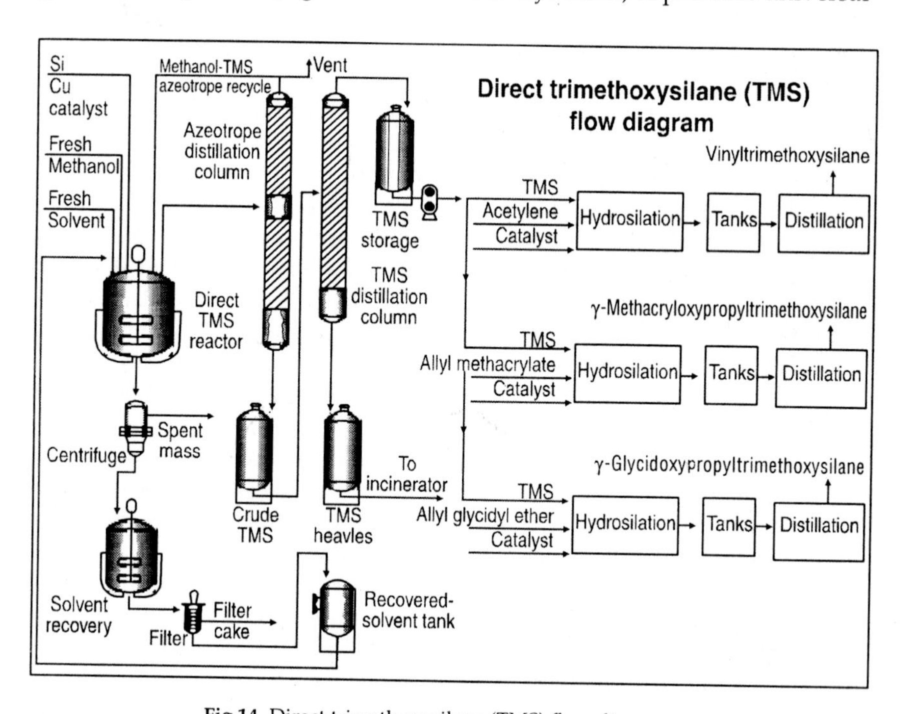

Fig.14. Direct trimethoxysilane (TMS) flow diagram.

raw material for synthesis of silicon-based end products, has so far not been commercially successful[882]. Above WITCO process for direct production of trimethoxysilane (TMS), recently (November 1999) received Kirkpatrick Chemical Engineering Achievement Award. Historically, this process was first developed by WITCO Corporation, at its Termoli (Italy) plant and later the pilot plant was erected at C.T. Witco Inc, Tarrytown, New York (USA) facility. This direct reaction between methanol and silicon, contrary to the conventional process of reacting alkylchloride or HCl and silicon, producing chlorosilane, subsequent esterification by alcohol and finally undergoing hydrosylilation reaction for adding organic functionality (which is inefficient, capital intensive and environment polluting), has fewer process steps, is low in capital requirement, do not emanate corrosive HCl or halogen wastes, having 75 % lower waste load, and has added flexibility like possibility of erecting new alkoxysilane capacity without the associated need for chlorosilane plant and the reactor operates with low hold-up, which allows it to stop and start anytime, thus minimizing the need for a large TMS inventory. The process as shown in Fig.14. is carried out in a well agitated slurry reactor that contains both the silicon metal and copper catalyst, while methanol is introduced as vapour and the reaction is carried out in a high boiling liquid hydrocarbon solvent. Starting and effectively running the reaction thus requires effective gas-liquid and gas-solid mass transfer. The vapour effluent from the reactor is fed to a distillation column, from top of which the lower boiling TMS-methanol azeotrope escapes, which is later condensed and returned to the reactor. The higher boiling stream consisting mainly of TMS as well as some tetramethoxysilane and higher components (but only 0.05-0.09 wt % methanol), goes down the stripping column and is removed for subsequent rectification. To avoid the reaction of excess methanol with product TMS, it is necessary to remove excess unreacted methanol from the mixture. But methanol and TMS forms a low boiling azeotrope whose normal boiling point 62.5°C is quiet close to methyldimethoxysilane (BP=61°C) and methanol (BP=64.5°C), but much lower than that of TMS itself (84°C). Thus crude product mix cannot be separated into pure methanol and TMS fraction by distillation alone but can be achieved by recycling part of TMS that is tied-up in azeotrope. Also, the system is likely to develop high pressure due to rapid phase change occurring due to superheating of methanol in contact with the solvent, which is comparatively nonvolatile and immiscible. This intermediate product (TMS), can further be converted by hydrosylilation reaction to commercial vinyl, epoxy and methacryloxy derivatives. These alkoxysilanes are essential components for fiberglass reinforced plastics, high durability coatings, low rolling resistant tyres and other high performance products which command a global market of over $500 million/yr.

On the other hand, market for pure silicon, as mentioned earlier is slowly seeing competition with other nonsilicon photocell materials and a mixed market is envisaged in the period 2000-2010, after which with technical progress a real spread of PV-residential system is expected to emerge[883]

CHAPTER 7

Environment and Health Aspect of Silicon and Its Compounds

As has been mentioned in the first chapter, silicon photocells produce electricity in most environment-friendly and non-polluting way. A real test to this effect was carried out recently (October 1996 to September 1997) by Watnabe and Haruyama[885], who installed a 3 kW photovoltaic system in a residential area on the roof of a house, which in the above period generated 3152 KWH, about 63 % of the load needed for the house, and it was found at the end of 1 year test period that it saved a good part of the electricity cost and reduced CO_2 emission in real sense, compared to conventional thermoelectric power generation. These silicon cells at the end of its life period, get mixed with its natural counterpart silica and thus do not cause any environment pollution.

As far as silicon-compounds are concerned, in general, all available literature indicates most of its compounds are harmless to human health as the polymeric silicon bonds are quite stable (Cleavage of Si-O and Si-C bond do not occur with our body fluid), nor do they coagulate blood or adhere to body tissue. Nevertheless, compared to all other chemical compounds, silicon-compounds constitute the greatest quantum of chemicals that are in daily use and constant study (especially long range toxicology) is a necessity to safeguard our health. Here we will discuss, some such findings with respect to direct use of silicon compounds (i.e. in contact with human body), its effect on other animals as it is being carried away by water and air, its overall effect on environment, and the effect of silicon industry to gross environment changes (e.g. global warming, ozone depletion, climate change, etc).

Chlorosilanes on hydrolysis releases hydrochloric acid and thus can irritate the skin, and some formulations which contain other substances along with silicon-compounds, also may cause skin allergy. Methyl-silicones, phenyl-polymers, alkyl and fluoroalkyl substituted polymers do not cause any toxic effect to human body. Fluoro compounds like trifluoropropylsilicone when heated to 280 °C, dissociates and emanates toxic gases[886]. Tetraorganosilanes have no toxicity, but organofunctional silanes have low level of toxicity[887]. Sometimes these effects are used to advantage for designing drugs. Silicon functional compounds also show some toxicity, especially when they have readily hydrolyzable group, which irritates mucous membranes and eye tissues. Silanes that contain both organofunctional and silicon functional groups, show toxicity to a greater extent. This is valid with almost all silatranes which contain both amino and alkoxy groups. PDMS is used by pharmaceutical companies for its inertness to warm blooded animals. Administration of PDMS in chronic oral doses to various animals did not reveal any harmful effect. However, implantation of silicon gels under human skin reported controversial results. Injection of PDMS forms tumour at injection point, showing rejection by body tissues. Inhalation of PDMS containing aerosols for 90 days to experimental animals, did not show any adverse effect. Methylchlorosilanes are inflammable and care should be taken during storage and transportation of the compound.

Polydimethylsiloxanes which constitute the major bulk of our silicon consumption, are non-biodegradable. However, these materials when carried away by water gets absorbed in sewage sludge (as they are of high molecular weight) and thus eventually gets eliminated. Siloxanes that find their way to soil, also get degraded by certain clay minerals. Low molecular weight siloxane polymers (oligomers), have relatively higher vapour pressure at room temperature and thus escape to atmosphere. In atmosphere they get photocatalytically degraded (half-life being few days only) to silicates. But effect of these extremely fine silicate particles on our body has not been evaluated yet. Solid forms of siloxanes, on the other hand, like rubbers and resins on burning in incinerators produce harmless water vapour, carbon dioxide and silica. Transportation of some silicon-compounds needs to be taken carefully, as many of them are inflammable and spilling may cause dangerous accumulation at a small area. Disposal of silicon industry by-products to sewerage and ultimately to river streams and seas may end up in marine animal bodies (fish), which in-turn are consumed by human beings. This establishes a part of silicon compounds used from day to day in a closed biological cycle and enhances chance of accumulation with time. Dr. H. Tveit of Elkeme Silicon Division, Norway presented a paper on overall environment pollution by silicon compounds (which includes global warming, ozone depletion, etc. and setting a new performance indicator in the light of recent Kyoto agreement on global environment pollution), at 'Silicon based Chemical Industries, IVth International Conference (1998)', and readers interested further on the subject may consult the above paper for further details.

CHAPTER 8

List of Silicon Manufacturers

8.1 PRODUCERS OF SEMICONDUCTOR GRADE SILICON

(a) India

Name of the producer	*Capacity*
(i) Mettur Chemical and Industrial Corporation	50 ton/year
(ii) Silatronics	10 ton/year

Besides the above silicon rod manufacturers, there are a number of industries who process the basic silicon rod into thin wafer cells as well as silicon photovoltaic modules. Some of these manufacturers are:

(i) Webel-Siemens Industries, Salt Lake, Calcutta (West Bengal).
(ii) Bharat Heavy Electricals Ltd, Bangalore (Karnataka State).
(iii) Central Electronics Ltd, Shahibabad (Uttar Pradesh).
(iv) Renewable Energy India Ltd, Jaipur (Rajasthan State).
(v) Udhaya Semiconductor Pvt Ltd.
(vi) Indian Metal and Carbides Ltd.
(vii) Suryavonics Ltd, Hyderabad (AP)
(viii) Metakem Silicon Ltd, Mettur-Dam (Tamil Nadu).
(ix) Suryanamo Technologies Ltd (in collaboration with Spire Corpn, USA), 539-540 Chandralok, Secunderabad 500 003, India.

Metakem Silicon Ltd, has designed and installed a 10 ton per year polysilicon production plant along with recycling facility. A few systems based on amorphous silicon photocell, have also been designed and manufactured by Central Electronics Ltd, Sahibabad, REIL (Jaipur), and the export oriented plant Suryavonics.

(b) USA

American semiconductor grade silicon manufacturing units are much larger in capacity (e.g. Monsanto plant produces to the tune of 2000 ton/

year polycrystalline silicon). List of the other silicon manufacturer (mostly via trichlorosilane route) are as follows:

Name of the producer	*Production route*
(i) Great Western (subsidiary of General Electric)	$TCS/SiCl_4$
(ii) Hemlock Semiconductor (subsidiary of Dow-Corning)	$TCS/SiCl_4$
(iii) Monsanto (Electronic Material Division)	$TCS/SiCl_4$
(iv) Motorola (Semiconductor Products Division)	$TCS/SiCl_4$
(v) Texas Instruments	$TCS/SiCl_4$
(vi) Union Carbide Corporation (Electronic Division)	SiH_4 via chlorosilanes

(c) Japan

Name of the producers	*Production route*
(i) Chisso Corporation	TCS
(ii) Osaka Titanium Corporation	TCS
(iii) Shin-Etsu Chemical Industries Ltd.	TCS
(iv) Komatsu Semiconductor Ltd.	SiH_4

(d) Germany

Name of producer	*Production route*
Wacker Chemetronics GmbH (1000 ton/year plant)	TCS

(e) Italy

Name of producer	*Production route*
Smile (Owned by Dynamite Nobez)	TCS

(f) Denmark

Name of producer	*Production route*
Topsil (Owned by Motorola)	$SiCl_4$

(g) Russia

Name of producer	*Production route*
Joint Stock Company, Lipetsk, Russia 399820	$TCS/SiCl_4$

8.2 PRODUCERS OF SILICON-COMPOUNDS

USA

Wacker USA (no silane production facility).
WITCO Corporation (Organosilicon Division).

Dow-Corning (USA).
Union Carbide Corporation.

Germany

Bayer A.G.
Goldschmidt (no silane production facility).
Wacker Chemetronics GmBh.

France

Rhone-Poulenc Ltd.

UK

Dow Corning U.K Ltd.

Norway

Elkem Silicon (7300 Orkangu).

Japan

Shin-Etsu Chemical Industries Ltd.
Torray Silicones.
Toshiba Silicones Ltd.
Tokuyama Soda Ltd.
Canon K.K Ltd.

Russia

Joint Stock Company 'Silane', Lipetsk, Russia 399820

India

Metrowork Silicon Ltd (Calcutta); High Temp/High Vacuum Grease main product. Silicones Industries (India) Ltd (Delhi); Surface coating for building materials. Dynatron India Ltd (in collaboration with Dynamit Nobel), 14-15 Old Court House St. Calcutta 1 (producer of industrial laminates & resins).

References

1. *E.M. Separated stable isotope price list:* Oak Ridge National Laboratory, Isotope distribution office, Oak Ridge, Tennessee (1987).
2. *Separated Isotope: Vital tool for science and medicine*—Newman E. National Academy Press, Washington D.C., pp 45 (1982).
3. *Inflatable silicone lenses for space program (Fresnel type lenses)*—O'neil M.J. and Piszczor M.F. (Entech Inc, Keller, Tx 76248, USA). Conf Rec IEEE Photovoltaic Spec Conf 26th, pp 853–856 (1997), published proceeding (1998).
4. *Fresnel lens for deep space solar collector (refractive liner element technology based)*—Murphy D.M., and Eskenazi M.I., (AEC-ABLE Engineering Inc, Golcta, CA 93117, USA); ibid, pp 861–864 (1997).
5. Development of metallic mirror for solar collector—Alimov A.K., Alvutdinov D.N., Gaziev U.K., Guner E.A., Rakhimov R.A. and Kamaldinov R.G. (Fizic Technical Inst, Russia). *Geliotekhnika*, 1, 39–43 (1997).
6. Silicon solar cells characteristics at high temperature—Bakirov M.Y. (Azerbaijan, Russia). *Geliotekhnika*, 1, 9–12 (1997).
7. *Handbook of Extractive Metallurgy* (Vol. IV)—Ed. Habashi F.; Publisher—Wiley—VCH, N.Y. (1997).
8. Light omission from Er-doped silicon—Coffa S., Franzo G. and Priolo F. *MRS-Bulletin*, 23 (4), 25–32 (1998).
9. Er-doped silicon for microphotonics and integrated optoelectronics—Fitzgerald E.A. and Kimerling L.C. (USA); ibid, 23(4), 39–47 (1998).
10. Metallic copper: An efficient catalyst with a wide range of silicon metal—Laroze G. and Bultean T. (Rhone Poulene Industrialization, 69192 Saint Fons, France). *Silicon Chem. Indus. IV Conf. Proc.*, Ed by Oeye H.A., pp 393–399 (1998).
11. *Action of copper on TCS synthesis*—Ehrich H., Lobreyer T., Hesse K. and Lieske H., ibid, pp 113–122 (1998).
12. *Thermodynamic and kinetic study of the Ni-Si-Cl-H system: Relevance of transition metal complex catalysts in TCS production*—Acker J., Bohmhammal K. and Roewer G., ibid., pp 133–143 (1998).
13. *Study on selectivity in TCS producing reactions*—Wakamatsu S., Hirota K. and Sakata K., ibid, pp 123–132 (1998).
14. *Manufacture of solar grade silicon by carbothermic reduction of silica*—Kato Y., Abe M., Hanazawa K., Baba H., Nakamura N., Yushita K. and Sakaguchi Y. (Kawasaki Steel Corpn., Japan). Jpn Kokkai Tokkyo Koho JP 10 265, 213 (October 8th, 1998).
15. Hollow electrode injection in silicon metal production—Larochelle P., Bosivert R. and Ksinsik D. *Silicon Chem Indus IV Conf Proc*—Ed by Oeye H.A., pp 33–40 (1998).
16. Supporting high productivity in silicon manufacture with prebaked electrode technology—Klotz J.A., Boardman T.R. and Boardwine C.E., ibid, pp 41–50 (1998).
17. Raw material for high quality silicon production and specific features of carbothermal reduction—Prokhorov A.M., Petrov G.N., Issamonov N.A. and Tkacheva T.M., ibid, pp 85–92 (1998).

18. High purity polysilicon crystal; Different processes, quality requirement and the market—Lobreyer T. and Hesse K. (Wacker-Cheme GmbH, D-84479 Burghausen, Germany). ibid, pp 93–100 (1998).
19. Refining and characterization of silicon for the chemical industry—Neto J.B.F., Nrgueira P.F., Kashiwaba J.J., Cristo S.C., Rodrigues D. and Pinto E.C.O. (Metallurgy Division, Inst of Tech Research, Sao Polo CEP 05508–901, Brazil); ibid, pp 101–104.
20. Silicon metal refining at Electrosilex plant (CEP 39445–000, Brazil)—Da Silveira R.C., Lamas M.S., Cysne M.A., Savini G., Braga H.C., Neto J.B.F., Norguiera P.F. and Oleveira R.A., ibid, pp 377–388 (1998).
21. *Graphite container for electron beam melting of silicon*—Hanazawa K., Abe M., Baba H., Nakamura N., Yuge K. and Kato Y., Jpn. Kokkai Tokkyo Koho J.P. 11 180, 712 (Appl. 19 Dec., 1997).
22. *Recovery of sulfuric acid and recycling of byproduct in phosphoric acid plant*—Harad I., Yusutake T. and Inoue H. Jpn Kokkai Tokkyo Koho JP 10 338,514 (22 Dec., 1998).
23. Manufacture of silicon-tetrafluoride from phosphoric acid plant—Kojima K. and Okuda T. (Aichi Steel Works, Japan). Jpn Kokkai Tokkyo Koho J P 09 183, 608 (15 July, 1997).
24. Sanjurjo, A. *et al.*, *J. Electrochem Soc.* 128, 179–181 (1981). L. Nanis *et al.*, Final report DOE/JPL—94, 4471, March (1980). *Silicon by reducing* SiF_4 *with Na.*
25. Production of high purity silicon for solar cell and electronic application by TCS process—Mazumder, B.; *Transac Indian Ceram Soc.*, 40(4), 155–159 (1981).
26. *Apparatus for manufacturing sulfur-trioxide by oxidation of sulfur-dioxide*—Nagaya Y., Yanagi H., Katagiri T., Nakamura Y. and Kodama T. (Hitachi Shipbuilding and Engineering Co Ltd. Japan). Jpn. Kokkai Tokkyo Koho JP 11 171, 509 (99,171,509), Appl. 4 December (1997).
27. *Sulfuric acid recovery apparatus*—Yoshioka T. and Horie K. (Ebara Corpn, Japan). ibid, JP 11 157, 813 (99,157,813), 28 November (1997).
28. *Manufacture of concentrated sulfuric-acid*, Dietrich K., Doni J. and Doerr K., Ger Offen D.E. 19,522,927 (2 Jan, 1997).
29. *Manufacture of trichlorosilane from dichlorosilane*—Oda H. Jpn. Kokkai Tokkyo Koho JP 11 29, 315 (2nd February, 1999).
30. *Recovering elemental silicon from reaction residue in a fluidbed reactor with hydrochloric acid and methyl-chloride to produce TCS*—Streussberger H., Streckel W., and Riedle R., Ger Offen 2,807,951 (30 August, 1979).
31. Production of TCS in a fluidized bed reactor—Li K.X., Peng S.H., and Ho T.C. *AIChE Symp. Ser.*, 84, 114–125 (1988).
32. *A process for manufacturing polycrystalline silicon with high yield*—Oda K. (Tokuyama K.K., Japan). Jpn. Kokkai Tokkyo Koho JP 11 49,508 (23 February, 1999).
33. *Manufacture of polycrystalline silicon containing small amount of carbon* —Oda. K. Jpn. Kokkai Tokkyo Koho J.P. 11 49,509 (23 February, 1999).
34. Production of silicon from rice husk—Acharya H.N., Banerji H.D. and Sen S. *Proc. National Solar Energy Convention*, IIT (Bombay), pp. (1979).
35. Solar grade silicon—Prasad N.K.S.; *International Solar Energy Congress, Extended Abstrc ISEC* (Vol. I), pp. 456; *Intl. Solar Energy Congress 1977 proc.*, New Delhi, India (published 1978).
36. Highly purified silane gas for advanced silicon semiconductor devices—Ohki A., Ohmi T., Date J. and Kijima T. (Dept. of Electronics Engineering, Tohoku University, Japan). *J. Electrochem Soc.* 145 (10), 3560–3659 (1998).
37. Manufacture of granular polysilicon by pyrolysis decomposition of silane—Liyama S. and Sokai J. (Tokuyamma K.K., Japan). Jpn. Kokkai Tokkyo Koho JP 10 158,006 (16 June, 1998).
38. A comparison between TCS and silane route in the purification of metallurgical grade silicon to semiconductor quality—Brenman W.C. and Dawson H.J., *Silicon Chem. Indus. IV Conf. Proc.*—Ed. by Oeye H.A., pp. 101–112 (1998).

39. *Production of silane compound by disproportionate reaction of halosilanes using a quartornary ammonium salt based catalyst*—Hattori N., Sokata K. and Kameda M. Jpn. Kokkai Tokkyo Koho JP 11 156, 1998 (Appl. 25 November, 1997).
40. *Manufacture of silane containing disilane in high yield* (Mitsui Toatso Chem., Japan). Jpn. Kokkai Tokkyo Koho JP 09 156,917 (17 June, 1997).
41. *Manufacture of monosilane by disproportionation of chlorosilane*—Sakata K. and Hirota K. (Tokuyama Soda Co., Japan). Jpn. Kokkai Tokkyo Koho JP 10 59,707 (3rd March, 1998).
42. Manufacture of high purity silicon in closed cycle—Natsume Y. and Kumanao H. (Osaka Titanium Co Ltd, Japan). ibid, JP 11 92,130 (99 92,130). (6 April, 1999).
43. Influence of Li^+, Na^+, Mg^+, Ca^{+2}, Sr^{+2}, Cu^{+2}, $A1^{+3}$, La^{+3} and Y^{+3} on silicagel formation by investigating hydrolysis and polycondensation of silicon-tetraethoxide—Alfa Products (Vanvers, MA, USA), *J Am Ceram Soc.*, 73(9), 264 (1990).
44. *Annual Report* of 'National Physical Laboratory', New Delhi 110 012, India (1997).
45. Properties and uses of colloidal silica—Dolezal J. (Silchem spol sro., Usti Nad Labem, Czech Republic 40331). *Chem. Magazine* 8(4), 14–16 (1998). in Czech.
46. *Process for making silanized colloidal silica*—Van V., Peter H. and Shirin W. (Dendreon Corpn, USA) PCT Intl Appl WO 99 36,359 (22nd July, 1999).
47. Particle sizes of silica—Barthel H., Heinemann M., Stintz M. and Wessely B. (Wacker-Chemie, Germany). *Chem Engg. Technol.*, 21(9), 745–752 (1998).
48. Silicon-nitride manufacture from silicon-tetrachloride—Joffers P.M. and Bauer S.H., *J. Non-cryst Solid*, 57(1), 189–193 (1983).
49. *Silicon-nitride whiskers*—Motojima S., Hattori T. and Ishikawa N. Jpn Kokkai Tokkyo Koho JP 02034 598 (1990).
50. *Manufacture of silicon-nitride powder in fluidizedbed reactor*—Konya Y., Sasagawa T., Watnabe M. and Fukuhira M. (Shin-Etsu Chemical Industries, Japan). Jpn Kokkai Tokkyo Koho JP 10 130,006 (19 May, 1998).
51. *Preparation of alpha-siliconnitride*—Kasai K. and Tsukidate T.; Jpn Kokkai Tokkyo Koho JP 79, 124, 898 (28 September, 1979).
52. Production of silicon-nitride power from silicon-tetrachloride and ammonia in arc plasma furnace—Perugini G.; *Conf. Proc. Intl. Symp Plasma Chem 4th*, 2, 779–785 (1979).
53. *Chemical vapour deposition technique for large diameter silicon production*—Keck D.W., Russell R.O. and Dawson H.J.; PCT Intl Appl WO 99 31,013 (appl 15 Decem. 1997).
54. Preparation of semiconductor grade silicon by decomposition of silane—F1. Niedam, Gh. Mozes and Grigorivici E.; *Revista de Chemie II*, 468–476 (1960); Translated by H. Mantsch, National Research Council of Canada (Technical Translation), Ottawa (1971).
55. Large grain polycrystalline silicon by CVD method with silicon-tetrachloride and hydrogen mixture—Beckloff B.N., Lackey W.J. and Pickering E.M.; *J. Mater Res.*, 14(3), 672–681. (1999).
56. Fundamental understanding and integration of rapid thermal processing (PECVD) and screen printing for cost effective, high efficiency silicon photovolatic devices—Doshi P.M. (Georgia Inst of Technol, Atlanta, USA). UMI Order No. DA 9810409 (283 pages), 1997. *Diss Abstr. Int. B*, 58(9), 5002(1998).
57. Fast deposited microcrystalline Silicon Solar Cells—Torres P., Meitet J., Kroll U., Beck N., Keppnet H., Shah A. and Malang U. Conf. Rec IEEE Photovoltaic Spec Conf. 26, 711–714 (1997).
58. Polycrystalline silicon thin film and solar cell prepared by rapid thermal CVD—Zhao Y., Jiang X., Wang W., Li Zh, Yu Y and Liao S. (Beijing Solar Energy Res Inst., Beijing 100083, People Rep. China); *Conf. Rec. IEEE Photovoltaic Spec. Conf.*, 26, 731–733 (1997).
59. High speed deposition of silicon film by PECVD from trichlorosilane—Rostalsky M., Kunze T., Linke N. and Muller J. ibid, pp 743–746 (1997).
60. Effect of hydrogen on silicon crystal quality in PECVD—Ito T, Imaizumi M., Konomi I. and Yamaguchi M. ibid, pp 747–750 (1997).

61. Highly stabilized amorphous silicon from SiH_2Cl_2 (CVD process)—Dairiki K., Yamada A., Kanagai M., Merkulov A. and Asomoza R. ibid, pp 779–804 (1997).
62. High rate deposition of microcrystalline silicon by PECVD under high pressure—Guo L., Kondo M., Furukawa M., Saitoh K. and Matsuda H. *Japan J Appl. Phys.* (Part II), 37 (10A), 1116–1118 (1998).
63. Production of silane—Brenemann W.C., Farrier E.G., and Morthara H. *Proc. 13th, IEEE Photovoltaic Spec Conf,* Wash D.C. (USA), pp. 339 (Pub. 1979).
64. New ways to make silicon crystalls for semiconductor use—Perkinson G., Ushio S., Short H., Hunter D. and Lewald R.; *Chem. Engg.*, May 26, pp 14–16 (1987).
65. Process for preparing high purity silicon granules—Schreider F. and Kim Hee Y. Eur Pat Appl. EP 896,952 (17 February, 1999).
66. Chemical grade silicon particles produced by atomization—Rodrigue D., Filho C.A.F.R., Fereira N.J.B., Kashiwaba J., Salgado L. and Noguiera P. *Mat. Sci. Forum* 299–300 (1999). Advance Powder Technology 182-189 (1999).
67. Rubin B., Moates G.H. and Weiner J.R. *J Electrochem Soc.* 104, 656 (1957); Japan Telegram and Telephone Co., JA 13669 (1988).
68. Manufacture of high purity silicon and the apparatus—Kendo J., Shimada H., Tokumaru S., Watnabe R., Nogasim A. and Iyose A. PCT Intl. Appl WO 99 33,749 (8 July, 1999).
69. *Si-F-H semiconductor material for solar cell*—Ovshinsky S.B.; Report of Energy Conversion Devices (USA), (1979).
70. *Purification of silicon to solar grade, apparatus for the process*—Nakamura N., Abe M., Hanazawa K., Baba H., Yuge K., Sakaguchi Y. and Kato Y. Jpn. Kokkai Tokkyo Koho JP 10 273,311 (13 October, 1998).
71. *Manufacture of polycrystalline silicon ingot for solar cell*—Yushita K., Nakamura N., Abe T., Hanazawa K., Baba H., Sakaguchi Y. and Watnabe S. ibid, 10 273,313 (13 October, 1998).
72. *Purification of silicon for solar cell*—Sakaguchi Y., Abe M., Baba H., Nakamura N. and Kato Y. ibid, JP 10 182,129 (7 July, 1998).
73. *Method for purifying silicon for solar cell*—Hanazawa K., Abe M., Baba H., Nakamura T., Yuge K., Sakaguchi Y. and Kato H.; ibid, JP 10 273,374 (13 October, 1998).
74. *Purification of silicon for solar cell by electron beam melting in graphite container*—Hanazawa K., Sakaguchi Y. and Kato Y. ibid, JP 10 182,133 (7 July, 1998).
75. *Purification of silicon for solar cell*—Hanazawa K., Sakaguchi Y. and Kato Y. ibid, JP 10 182,130 (7th July, 1998).
76. *Stirring of molten silicon during purification of silicon by electron beam melting*—Hanazawa K., Yushita K. and Kato Y. ibid, 10 182,138 (7 July, 1998).
77. *Electron beam melting of silicon for purification of silicon*—Hanazawa K., Akawa K., Satio A., Ikeda T. and Kato Y. ibid, 10 182,131 (7th July, 1998).
78. *Purification of silicon by unidirectional solidification of molten silicon for solar cell making*—Sakakuch Y., Abe M. and Kato Y. ibid, JP 10 182,135 (7 July, 1998).
79. *Purification of silicon for solar cells*—Hanazawa K., Sakaguchi Y. and Kato Y. ibid, JP 10 182,130 (7th July, 1998).
80. *Oxidative removal of boron from silicon in solar grade silicon*—Baba H., Nakamura N., Abe M., Sakaguchi Y. and Kato Y. ibid, JP 10 182,136 (7 July, 1998).
81. *Method and apparatus for purification of silicon by solidification for solar cells*—Sakaguchi Y., Yushita K. and Abe M.; ibid, JP 10 182,137 (7 July, 1998).
82. *Manufacture of polycrystalline silicon ingot for solar cell*—Baba H., Nakamura N., Abe M. and Sakaguchi Y. ibid, 10 194,718 (28 July, 1998).
83. *Manufacture of silicon by melt reduction*—Tamura M. ibid, JP 10 212,111 (11 August, 1998).
84. *Manufacture of high purity silicon powder*—Shugie J. and Kawakami H. (Toyota Motor Corpn., Adoma Techa K.K., Shin-Etsu Chemical Industries, Japan). ibid, JP 10 182,128 (7 July, 1998).

85. Fused salt electrodeposition of thin layer silicon—Moore J.T., Wang T.H., Heben M.J., Douglas K. and Ciszek T.F.; *Conf Rec. IEEE Photovoltaic Spec Conf 26th, Proc.*, pp 775–778 (1997).
86. *Hydrolysis method of producing pure silicon for solar cells*—Kon Y., Kawasaki N. and Nakamura K. (Nippon Steel, Japan). Jpn. Kokkai Tokkyo Koho, JP 11 189,408 (13 July, 1999).
87. *Process for electrolytic production of metal in molten state*—Ginatta M.V. PCT Intl. Appl WO 98 33,956 (6 August, 1998).
88. *Silicon for solar cell*—Ciszek T.F. (SERI, Golden, Co., USA). *Crystal Growth of Electronic Materials*—Ed. by Kaldis E., pp 185–210 (Elsevier Pub), (1985).
89. Solidification of polycrystalline silicon ingots: Comparison between conventional casting and electromagnetic casting process—Ehret E. and Langier S.; *Conf Rec IEEE Photovoltaic Spec Conf 25th*, Arlington (VA, USA), pp 613–616 (1996).
90. *Process and apparatus for manufacture of polycrystalline silicon and silicon-substrate for solar cell*—Avatani F., Kato Y., Sakaguchi Y., Yuge N., Baba H., Nakamura N. and Hanazawa K. PCT Intl. Appl WO 98 16,466 (23 April, 1998).
91. Westinghouse process for making dendritic silicon cell—*The Engineers*, issue dated 3rd April, 1980.
92. Crystallization of silicon on insulators (Part I & II) in *Laser solid interaction and transient thermal processing of materials*—ed by Narayan W.L., Brown B., and Lemons R.A.; *Proc. 1982 Mat Res Soc Conf*, Boston, USA; North Holland Pub., N.Y., pp 463 (1983).
93. Ibid (Part IV)—ed. by Appleton B.R. and Celler G.K., pp 459 (1983).
94. Ibid. (Part IV)—ed. by Gibbons J.F., Hess L.D. and Sigmon T.W., pp 413 (1983).
95. Scanned microzone crystallization to form single crystal silicon on amorphous insulator; a review—Sedgwick T.O.; *Crystal Growth of Electronic Materials* (Elsevier), pp 229–244 (1985).
96. Substrate for thin silicon solar cell—Blakers A.W.; *Solar Energy Mat. Sol. Cell*, 51(3–4), 385–392 (1998).
97. Low pressure CVD of polycrystalline silicon for growing thin film silicon on glass substrate—Bergmann R.B., Kiruke J., Strunk H.P. and Werner J.H. (Max-Planck Inst., Stuttgart, Germany). *Mater Res Soc Symp Proc*, 325–330. (1997).
98. CVD deposition of crystalline silicon on glass for thin film solar cell—Bergmann R.B, Brendel R., Wolf M., Loelgen P., Werner J.H., Krinke J. and Strunk H.P. *Conf Rec IEEE Photovoltaic Spec Conf 25th*, Arlington (VA, USA), pp 365–370 (1996).
99. *Process for forming thin film silicon SOI solar cell and area sensors*—Iwane M., Yoneshara T. and Ohmi K. (Cannon Kabushiki Kaisa, Japan). Eur Pat Appl EP 867,923 (CI HOIL 21/304) (30 September, 1998).
100. Polycrystalline silicon film formation on foreign substrate by rapid thermal CVD technique—Slaoui A., Monna R., Angermeir D., Bourdais S. and Muller J.C.; *Conf Rec IEEE Photovoltaic Spec Conf 26th*, pp 627–630 (1997).
101. Improvement of large area SCAF structure amorphous silicon solar cells with plastic film substrate—Tabuchi K., Fujikaka S., Sato H., Saito S. and Takno A. ibid., pp 611–614 (1997).
102. *Manufacture of polycrystalline silicon solar cell on cheap substrate*—Nishid S. (Cannon K.K., Japan). Eur Pat Apple EP 831, 539 (25 March, 1998).
103. *Recrystallised and epitaxially thickened polycrystalline silicon layer on graphite substrate*—Kanze T., Hauttmann S., Seekamp J. and Muller J. ibid, 735–738 (1997).
104. *Improved performance of thin film silicon solar cell on graphite substrate*—Auer R., Zettner J., Krinke J., Schultz M., Strunk H., Polisski G., Hierl T. and Hezel K. ibid, pp 739–742 (1997).
105. *Faster deposition of amorphous silica using gas jet on a substrate*—Jones S.J., Myatt A., Ovshinsky H., Doehler J., Izu M., Banerji A., Yang J. and Guha S. ibid, pp 659–662 (1997).

106. Manufacture of high purity silicon powder with uniform particle size—Abe M., Nakamura N., Baba H., Sakaguchi Y. and Kato Y. Jpn. Kokkai Tokkyo Koho JP 10 182,125 (7 July, 1998).
107. Plasma torch for purification of silicon—Abe M., Nakamura N., Babe H., Sakakuchi Y. and Kato Y. ibid, JP 10 182,127 (7 July, 1998).
108. Application of hot wire CVD: Electronic properties & device—Middya A.R., Guilet J., Brenot R., Perrin J., Bouree J.E., Longeuad C. and Kleider J.P. (CNRS, France). *Mater Res Soc Symp Proc*, pp 271–282 (1997).
109. Amorphous silicon deposition by HWCVD technique: How to exceed efficiency of amorphous silicon by 10%—Bauer S., Herbst W., Schroeder B. and Oechsner H. *Conf Rec IEEE Photovoltaic Spec Conf 26th*, pp 719–722 (1997).
110. Thin substrate based crystalline silicon solar cells with no grid shading—Aiken D.J. and Bernett A.M. *Conf Rec IEEE Photovoltaic Spec Conf 26th*, pp 763–766 (1997).
111. Development of light trapped, interconnected silicon film—Ford D.H., Rand J.A., Barnett A.M., Delledonne E.J., Ingram A.E. and Hall R.B. ibid, 631–634 (1997).
112. Study on SiON/SiN double layer anti-reflective coating on silicon solar cell–Quin J., Yang and Fu *J. Taiyangneng Xuebao*, 18(3), 302–306 (1997).
113. Honeycomb textured silicon solar cells—Zhao J., Wang A., Green M.A. and Ferazza F., *Appl. Phys. Lette.*, 73(14), 1991–1993 (1998).
114. Low thermal budget treatment of porous silicon surface on crytalline silicon solar cell: improved surfce passivation—Stalmans L., Poormans J., Bender H., Conard T., Jins S., Nijs J., Martens R., Strehlke S., LevyClement C., Debarge L. and Slaoui A. (Belgium). *Solar Energy Mat. Sol. Cell*, 58(3), 237–252 (1999).
115. Potential of amorphous silicon for solar cell: A review—Rech B. and Wagner H. (Inst. Schicht—Iontechnik, Germany). *Appl Phys. A Mater Sci. Process*, 69 (2), 155-167 (1999).
116. Crystal silicon thin film for solar cell: A review—Bergman R.B. (Inst. Physical Electornics, University of Stuttgart, Germany). ibid, 69(2), 187–194 (1999).
117. Separation of silicon isotopes by rectification of silicon tetrachloride—Orlov V.Y., Zhavononkov N.M.; *J. Appl. Chem. of USSR (Zh. Priki Ikhim)*, 29, 959–963 (1956).
118. Separation of boron isotopes—Green M. and Martin G.R., *Trans Farday Soc.*, 48, 416 (1952).
119. Separation of carbon isotopes—Baertschi P., Kuhu W., and Kuhu H.; Nature, 171, 1018 (1953).
120. Separation of silicon isotopes from silane—Devyatkh G.G., Borisov G.K., and Pavlov A.M.; *Dokl Akad Nauk SSSR* (English Translation), 138(2), 402 (1961).
121. Separation of silicon isotopes by distillation of methyl-silane—Brunken V.R. and Jost W., *Nachr Akad Wiss Gottingen*, 4, 123–126 (1962).
122. Silicon isotope separation by distillation of silicon-tetrafluoride—Mills T.R.; *Separation Science and Technology*, 25(3), 335–345 (1990).
123. Isotope effect of copper in ligand exchange system and electron exchange system observed by ion-exchange displacement chromatography—Martin A. Md., Nomura M., Fuji Y. and Chen J. *Separation Science & Technology*, 33(8), 1075–1087 (1998).
124. Separation of lithium isotope by ion-exchange chromatography—Taylor T.I. and Urey H.C., *J. Chem Phys*, 6, 429 (1938).
125. Separation of Lithium isotope by ion-exchange chromatography—Piez K.A., and Eagle H., *J. Am. Chem. Soc.*, 78, 5284 (1956).
126. Isotopic separation: Theoretical calculations—Bigeleisen J. and Mayer M.G., *J. Chem. Phys.*, 15, 261 (1947).
127. Contribution of nuclear size and shape, nuclear mass and nuclear spin towards enrichment factor of zinc isotopes in chemical exchange reactions by a cryptand—Nishizawa K., Yohisa M., Kawashiro F., Fuji T., and Yamamoto T. *Separation Science and Technology*, 33(14), 2101–2112 (1998).

128. Chemical reactions in molten salts (Part 21): Titanium mediated synthesis of silanes in chloroaluminate melts—Liesenhoff H., Sundermeyer W. (Anorganisch Chemische Institute, University of Heidelberg, D-69120 Heidelberg, Germany). *Z. Naturforsch B. Chem. Sci.*, 54(5), 573–576 (1999). German.
129. Marlett E.M. and DePriest R.N. (Ethyl Corpn.), U.S. 4778668 (1988).
130. Friedrich K.; Thesis R.W.T.H.—Aachen, Germany (1964).
131. Exxon Research and Engineering (Batelle Dev.), U.S. 3990,953 (1976).
132. Reaction mechanism of nucleophilic attack at silicon—Bassindale A.R., Glynn S.T., and Taylor P.G. *Chem. Org. Silicon Compd* 2 (Part I), 495–511 (1998).
133. Silatranes and their tricyclic analogue—Pestunovich V., Kiripichenko S., and Voronkov M.; ibid, pp 1447–1537 (1998).
134. Flory P.J., Crescenzi V. and Mark J.E., *J Am Chem. Soc.*, 86,146 (1964).
135. *An introduction to the chemistry of silicones*—Rochow E.G., John Wiley and Sons Inc., New York (1962).
136. *Inorganic polymers*—Barry A.J. and Beck H.N. (Ed by Stones FGA and Graham WAG), Academic Press Inc., N.Y. (1962).
137. *Organosilicon compounds*—Eaborn C., Buttersworth Scientific Publication, London (1962).
138. Organosilicon compounds—Bazant V., Chvalovsky V. and Rathousky J. Academic Press Inc., New York (1965).
139. *Stereochemistry, mechanism, and silicon*—Sommer L.H., Pergamon Press, New York (1962).
140. *Reaction heats and bond strength*—Mortimer C.T.; Pergamon Press, New York (1962).
141. Kirk-Othmer Chemical Encyclopedia (Vol 20). John Wiley & Sons, N.Y. (Publen 1980).
142. Different fluoride anion sources and trifluoromethylsilane: The first penta-coordinated silicon species with five Si-C bonds—Kolomeitsev A., MovchunV., Rusanov E., Bissky G., Lork E., Roschenthaller G.V. and Kirsch P., *Chem Commun.* (Cambridge), 11, 1017–1018 (1999).
143. Hypervalent silicon compounds—Kost D. and Kalikhman I., *Chem. Org. Silicon Compd.* 2 (Part-2), 1339–1445 (1998).
144. Structure and reactivity of hypercoordinated silicon species—Chut C., Corriu R.J.P. and Reye C. *Chem. Hypervalent Compound*—Ed. by Akiba K., Wiley-V.C.H. Publication, New York, pp 147–169 (1999).
145. Organosilicon derivative of fullerene—Ando W. and Kusukawa T. *Chem. Org. Silicon Compound* (Pt. III)2, 1929–1960. (1998).
146. Drake J.E. and Westwood N.P.C., *Chem. Ind. (London)*, 24, 112 (1969).
147. *Production of silicon-peroxide for oxidants, bleaches and disinfectant among other applications*—Koenigstein K.; Ger Offen DEI 9 714, 440 (15 Oct., 1998).
148. *Manufacture of hollow spherical aluminosilicate clusters in high yield*—Ooheshi F., Wada S. and Tsunofuji Y. Jpn. Kokkai Tokkyo Koho, JP10 236 818 (Sept, 1998).
149. Technical Brochure from Dynamite Nobez, Italy (1980).
150. Lopata S.L.; U.S. Patent 3,056,684 (11 October, 1962).
151. Berger D.M.; *Met Finish*, 72(4), 27 (1979).
152. Bahney R.H. and Harris L.A. (Dow Corning, USA). U.S. Pat. 4,197,230 (April, 1980).
153. Gagnon D. (Owens, Ill, USA). U.S. Pat. 4,103,065 (July 25, 1978).
154. Silane coupling agent used in solid rocket motor charges and their action mechanism—Zhao F., Shan W. and Li S. (Xian Modern Chem. Res. Inst.; Xian; People Rep. China). *Hanneng Cailao*, 6(1), 37–42 (1998).
155. *Preparation of haloorganosilane compounds*—Tsuchiya K. and Yamanchi K., Japan Kokkai Tokkyo Koho, JP11 199,588 (27 July, 1999).
156. Kamiya K. and Sakka S., *J. Mater. Sci.*, 15, 2937 (1980).

157. Kuznetov A.I.; USSR Patent No. 715,460 (1980).
158. *Preparation of fluorine containing chlorosilane by catalytic hydrosilylation in presence of platinum complex*—Tanaka A., Yamato Y. and Tsuchiya K. (Chisso Corpn, Japan). Jpn. Kokkai Tokkyo Koho, JP11 189,597 (13 July, 1999).
159. Krantz K.W. (General Electric, USA). U.S. Pat. 2,916,461 (Dec. 8, 1959).
160. Polmanteer K.E., Chapman H.L. and Lutz F.A.; *Rubber Chem. Technol.*, 58, 939–944 (1958).
161. Helbert J.N. and Saha N. J. *Adhesive Sci. & Technol.*, 9,905–925 (1991).
162. Scott R.N. Knollmeeller K. and coworkers; *Ind. Engg. Prod. Res. Dev.*, 19, 6 (1980).
163. Dunster A.M., Parsonage J.R. and Vidgeon E.A.; *Mat Sci. Technol.*, 5, 708–713 (1989).
164. *The analytical chemistry of silicones*—Smith A.L. (Ed); J. Wiley & Sons, N.Y. (1991).
165. Pollack J.P. and Fish J.G. (Texas Instruments, USA); U.S. Pat. 3,915,766 (28 Oct, 1975).
166. Growth & transport of structure controlled hydrogenated silicon clusters for deposition on solid surfaces—Watnabe M.O. and Kanayama T. *Appl Phys. (A), Mater Sci. Process*, A66 (Suppl Pt II), (1998).
167. *Preparation of hydrolyzable silyl group containing charge transporting agents for electronic devices*—Ishi R. and Nukata K. Jpn Kokkai Tokkyo Koho, JP11 124,387 (11 May, 1999).
168. *Silanyl triazines as light screening compound*—Huber U. Eur Pat. Appl EP 933,376 (4 August, 1999).
169. Voronkov M.G., Dyakov V.M. and Kirpichenko S.V., *J. Organomet Chem.*, 233, 1–147 (1982).
170. Tacke R. and Linoh H. in *The chemistry of organic silicon compounds*—Patai S. & Rappaport Z. (Ed), two vol.; Wiley Interscience, Chichester, U.K. (1989).
171. Moberk W.K., Baker D.R. and Feyenes G.J., *ACS Symp. Series 443*, 1–14 (1991); also U.S. Patent (to Dupont) 4510,136 (1983).
172. Hoechst, EP-A 224024 (1986) (H.H. Schubert *et al.*). Dainippon, JP-Kokkai 03(91)990003 (1989), (Sugamoto K. *et al.*), *Chem. Abstr.*, 115, 201146 (1991).
173. Chemically modified surfaces (Vol I): Silane, surfaces and interfaces—White W.C., Getting R.L. (Leyden Ed); Gordon Breach Publcn (London), 107–140 (1986).
174. Synthesis and fungicidal activity of S-allyl-o-substiuted phenylthiophosphates (-onates)-He Z., Liu J., Zhu Z. and Tang C.; *Gaodeng Xuexiao Huaxue Xuebao*, 20(2), 227–231 (1999).
175. Kilbourne F.L., Doede C.M. and Stasiunas K.J., *Rubber World*, 132, 193–197 (1955).
176. Borisov S.N., *Plaste und Kantschuk*, 10, 400 (1963).
177. Pierce O.R., Holbrook G.W., Johannson O.K., Saylor J.C. and Brown E.D., *Ind. Engg. Chem.*, 52, 783 (1960).
178. Rhone Poulac; French Patent 1188495 (12 July, 1957).
179. Brown J.F. (General Electric, USA). Belg Pat. 577012 (U.S. Prior), 24 March, (1958).
180. Peter J., Knope H. and Noll W. (Farbenfabriken Bayer). German Patent 1003,441 (12 November, 1953).
181. Pierce O.R., Holbrook G.W., Johannson O.K., Savior J.C. and Brown E.D. *Ind. Engg. Chem.*, 52,783 (1960).
182. Pike R.M. and Morehouse E.L. (Union Carbide); Ger. Appl. 1114326 (U.S. Prior) 12 October 1956.
183. Glaser M.A.; *Ind. Engg. Chem.*, 46, 2334 (1954).
184. Manufacture and application of lubricating greases—Boner C.J., Reinhold Publication, New York (1954).
185. *Mineralole und verwandte produkte*—Schultze G.R., Schmierfetle in C. Zerbe, Spinger-Verlag Publication, Berlin, Germany (1952).
186. Maxson M.T. and Lee C.L.; *Gummy Faseern Kunstst*, 39, 532–539 (1985).
187. Cornelius D.J. and Monroe C.M.; *Polym. Engg. Sci.* 25(8), 467–473 (1985).
188. Aminabhavi T.M. and Cassidy P.E.; *Rubber Chem. Technol.*, 63, 451–471 (1990).

189. Barry A.J. (Dow Chemical); U.S. Patent 2469625 (10 December, 1943).
190. Cheronus N.D. U.S. Patent 2564674 (7 August, 1947); U.S. Pat. 2579416 (25 Jan., 1946); U.S. Pat. 2579417 (7 August, 1947).
191. Cheronus N.D., U.S. Pat. 2635059 (26 October, 1948).
192. Cheronus N.D., U.S. Pat. 2579418 (24 January, 1949).
193. Cheronus N.D., U.S. Pat. 2568384 (13 November, 1947).
194. Goldschmidt Th.; French Patent 1298127 (German Prior 29 August, 1960).
195. Kauppi T.A. (Dow Corning); U.S. Pat. 2678893 (21 February, 1951).
196. Gilbert P.T. Jr.; Belg Patent 561185 (U.S. Prior 28 September, 1956).
197. Heyden R. and Plapper J., Bohme Fettchemie, Swiss Pat. 362483 (Germ Prior 14 May, 1959).
198. Ender H.H. (Union Carbide); U.S. Pat. 3078293 (23 September, 1960).
199. Alsgaard R. and Gilkey J.W. (Dow Corning), U.S. Pat. 2970881 (15 Dec., 1958).
200. Norton F.J. (General Electric), U.S. Pat. 2412470 (22 February, 1943).
201. Patnode W.J. and Norton F.J. (Thomson-Houston), French Pat. 948923 (U.S. Prior 16 November, 1940 and 30 July, 1942).
202. Elliott J.R. and Kriebel R.H. (General Electric), U.S. Pat. 2507200 (10 Feb 1945).
203. Haber C.P. (General Electric), U.S. Pat. 2553314 (1 July, 1944).
204. Barry A.J. (Dow Chemical), U.S. Pat. 2405988 (10 December 1943).
205. Kunzel H., *Dtsch Bauztschr*, 7,449 (1959).
206. Keil J.W. (Dow Corning). Germ. Appl. 1057442 (U.S. Prior 22 December 1955).
207. Leavitt H.J. (General Electric), U.S. Pat. 2985545 & 2985546 (26 March 1958).
208. Thomson J.F., *Paper, Film & Foil Converter*, 28, 22 (1954).
209. Collings W.R. (Dow Corning), Germ. Pat. 849737 (French Prior 29 March, 1949).
210. Reinke L.H. (Dow Corning), U.S. Pat. 3046155 (6 October 1960).
211. Baer M.E. and Concannon L.E., U.S. Pat. 2584413 (30 June, 1948).
212. Cooke H.H. and Russo D. (Standard Oil): U.S. Pat. 2526870 (5 December 1949).
213. Curie C.C. (Dow Corning); U.S. Pat. 2523281 (26 September, 1949).
214. Geen H.C., Quist J.D. and Johnson H.C., (Slimoniz); U.S. Pat 2812263 (10 March, 1949).
215. Swanson R.G. (DuPont); U.S. Pat. 2614049 (27 June, 1949).
216. *Inorganic polymers*—Bary A.J. and Beck H.N. (Stone FGA Ed.), Academic Press , New York (1951).
217. Brown E.D. Jr., *(Transac) Amer Soc. Lubric. Engg.*, 9,31 (1966).
218. Cupper R.A. and Shiffler R.W. (Union Carbide Corpn.), U.S. Pat 4,097,393 (21 June, 1978).
219. Polmanteer K.E. and Hunter M.J., *J. Appl Polym. Sci*, 1,3, (1959).
220. Koroleva T.V., Krasovskaya T.A., Sobolevskii M.V., Gornets L.V. and Ruskin Y.E. *Sov. Plastics*, pp 28 (January 1967).
221. Yerrick K.B. and Beck H.N.; *Rubber. Chem. Technol.*, 37,261 (1964).
222. Schiefer H.M. and Vandyke J., *(Transac) Amer Soc. Lubric Engg.*, 7,32 (1964).
223. Schiefer H.M., ibid, 9,36 (1966).
224. Borisov S.N. and Katlin A.V.; Sov. *Rubber Technol.*, 21(12), 4 (1962).
225. Borisov S.N., *Plaste und Kantschuk*, 10,400 (1963).
226. Harper J.R., Chapman A.D., Konkle G.M. and Khil J.W. (Dow Corning); Brit Pat. 859284 (U.S. Prior 15 November, 1957).
227. Wick M. *Kunststaffe*, 50, 433 (1960).
228. Adrianov K. *Proc. Akad Sci. USSR, Sect. Chem.* (English), 151, 616 (1963).
229. Merker R.L. and Scott M.J, *J Polym. Sci* 2,15 (1964).
230. Merker R.L., Scott M.J. and Haberland G.G., ibid, (A)2,31 (1964).
231. Pmietanski G.M. and Reid W.G., *VIth Joint Army-Navy-Airforce Conf. on Elastomer Research & Development*, Boston (USA), 18–20 October (1960).

232. Kniege W. and Schnurbusch K. (Farbenfabriken Bayer), Belg. Pat. 623408 (Germ Prior 13 October, 1961).
233. Nitzsche S. and Pirson E. (Wacker Chemie), Germ. Pat. 1068841 (21 June, 1956).
234. Swiss J. and Arntzem C.E. (Westinghouse Corpn), U.S. Pat 2595727, 2595728–2595730 (9 March, 1945).
235. Hurd D.T., Osthoff R.C. and Roedel G.F. *Industrial Engg. Chem.*, 40, 2078 (1948).
236. Lewis D.W., *Electrical Manufacturing*, 141, 360 (March 1957).
237. Weyenberg D.R. (Dow-Corning), U.S. Pat 2714099 (11 January 1954).
238. Patnode W.I. (General Electric), U.S. Pat 2503919 (11 January 1947).
239. Tyler L.J., Germ Appl. 1165867 (U.S. Prior 11 January, 1960).
240. Effect of alkyl group on structure and thermal stability of substituted lithiofluorosilylenoids—Feng D.C., Feng S.Y. and Deng. C.H., *Gaodeng Xuexiao Huaxue Xuebao*, 19(3), 451–454 (1998).
241. Hydrosilylation of carboxylic acid allylimides—Pashmova I.V., Trofimov A.E., Svetlichnays V.M., *et al. Rus J. Gen. Chem.*, 67(7), 1110–1115 (1997).
242. Electronic and steric effect of various silyl groups in radical addition reactions—Hwu J.R., King K.Y., Wu I.F., and Hakimelahi G.H. *Tetrahedron Lett.*, 39(22), 3721–3724 (1998).
243. The excited electronic states of silylidene—Hillard R.K. and Grev R.S. *J. Chem. Phys.*, 107(21), 8823–8828 (1997).
244. Electron energy loss and DFT/SCI study on the singlet and triplet excited states as well as electron attachment energies of tetramethylsilane, hexamethyldisilane, tris (trimethylsilyl) silane, and tetramethoxysilane—Huber V., Asmis K.R., Sergentra A.C., Allan M. and Grimme S. *J. Phys. Chem. A.*, 102(20), 3524–3531 (1998).
245. Lentz C.W., *Inorg. Chem.*, 3, 574 (1964).
246. Pines A.N. and Zinetek E.A. (Union Carbide Corpn); U.S. Pat 3,198820 (3 August, 1965).
247. Kahn F.J., Taylor G.N. and Schonborn H., *Proc. IEEE*, 61, 823 (1973).
248. Leyden D.E. and Luttrell G.H.; *Anal Chem.*, 47, 1612 (1975).
249. Leyden D.E., Luttrell G.H., Sloan A.E., and DeAngelis N.J. *Anal Chim Acta*, 84, 97 (1976).
250. Plueddmann E.P. (Dow Corning); U.S. Pat 4,071,546 (31 January, 1978).
251. Moses P.R. and Murray R.W. *J. Am Chem. Soc.*, 98, 7435 (1976).
252. Armstrong N.R. (Uniliver); Netherland Pat Apple 6,401,297 (Aug 17, 1964).
253. Isquith A.J., Abbott E.A. and Walters P.A.; *Appl. Microbiol.*, 23, 859 (1973).
254. Grohmann P. and Grohmann K., *Tetrahedron Lett*, 28, 2633 (1971).
255. Machleidt W., *Proc Internatl Conf Solid Phase Methods in Protein Sequence Anal.*, 17 (1975).
256. Lynn M. in *Inorganic support intermediates: Covalent coupling of enzymes on inorganic support*—Weetall H. (Ed), Marcel Dekker Publcn., New York (1975).
257. Synthesis of silicon containing alkenes by catalytic hydrosilylation of cyclopropylmethylene-cyclopropanes with triethylsilane—Bessmertnykh A.G., Donskaya N.A., Tveritinova E.V. and Beltskaya I.P. *Russ J.Org,Chem.*, 34(10), 1419–1422 (1998).
258. B $(C_6F_5)_3$-catalysed silation of alcohols: A mild general method for synthesis of silylethers—Blackwell J., Foster K., Back V.H. and Piers W.E. *J.Org Chem.*, 64 (13), 4887–4892 (1999).
259. Rhodium catalysed dominosilyl formylation of enynes involving carbocyclization—Fukuta Y., Matsuda I. and Itoti K. *Tetrahedron Lett*, 40(25), 4703–4706 (1999).
260. Transition metal silyl complex—Eisen M.S., *Chem. Org. Silicon Compd.*, 2 (Pt. 3), 2037–2138 (1998).

261. Steric effect of silyl groups—Hwu RJR, Tsay S.C. and Cheng B.L.; *Chem. Org. Silicon Comp.*, 2 (Pt. I), 431–494 (1998).
262. Alkali and alkaline—earth silyl compounds (preparation and structure)—Belzner J. and Dehnert U.; ibid, 2 (Part I), 779–825 (1998).
263. The hydrosilylation of propylene—Chernysev E.A., Belyakova Z.V., Yagodina L.A., Nikitinskii E.V. and Bykovchenko V.G. *Russ. Chem. Bull.*, 47(10), 1992–1995 (1998).
264. 1, 4-Silyl migration from oxygen to carbon in silyl-allyl ethers (kinetic and thermodynamic factors)—Mitchell T. and Schulze M. *Tetrahedron*, 55(5), 1285–1294 (1999).
265. Synthesis of bis-trimethyl silylated hydroxy alkynes—Bolourtchian M., Zadnard R. and Saidi M.R. *Indian J. Chem. (Sec-B) Organic Chem including Med. Chem.*, 37B (11), 1171–1173 (1998).
266. Cycloaddition of thermolytically generated methoxymethylsilylene to alpha, beta-unsaturated ketones and imines—Gehrus B., Heinicke J. and Meinel S. *Main Group Met Chem.*, 21(2), 99–104 (1998).
267. Iminosilanes as precursors for monomeric bis (silyl) aminomethylsilane and silylium-trichloroaluminate—Niesmann J., Klingebiel U., Ropken C., Noltemeyer M. and Herbst I.R. ibid, 2(4), 297–308.
268. Addition of trimethylsilylcyanide to aromatic ketones promoted by organic solution of lithium-salt—Jenner G., *Tetrahedron Lett.*, 40(3), 491–494 (1999).
269. One pot synthesis of silyl substituted dihydro-2H-pyran-2-ols from 3-phenyl-dimethyl-silyl-1-diethylamino-propyne and epoxides-Schabbert S. and Schaumann E. *Tetrahedron*, 55(5), 1271–1276 (1999).
270. A new silyl substituted arachno-tetraborane (8) derivative—Wesemann L., Ramjuie Y. and Wagner T. *Inorg. Chem. Commun.*, 1(11), 443–445 (1998).
271. Silylation reaction of olefins with monosilane and disilane in the presence of transition metal complexes, metal hydride, and radical initiator—Itoh M., Iwata K. and Kobayashi M. *J. Organometallic Chem*, 574(2), 244–245 (1999).
272. Hydrosilylation of cyclopropyl substituted methylenecyclopropanes with triethylsilane catalyzed by Wilkinson's complex—Bessemertnykh A.G., Grishin Y.K., Donskaya N.A., Tvertinova E.V. and Beletskaya I.P. *Russ. J. Org. Chem.*, 34(6) 790–798 (1998).
273. Convenient preparation method for beta-silylated olefins by hydrosilylation of methylenecyclopropane over rhodium-complexes—Bessmertnykh A.G., Blinov K.A., Grishin Y.K., Donskaya N.A., Tvertinova E.V. and Beletskaya I.P. ibid, 34(6), 799–808 (1998).
274. Synthesis and hydrosilyation (catalytic) of unsaturated boric–acid esters-Veliev M.G., Agaguseinova M.M., Mirzoeva D.O., Shatorova M.I., and Movshumzade E.M. ibid, 68(8), 1231–1234 (1998).
275. Trichlorosilylation of chlorogermanes and chlorostannes with $HSiCl_3/NEt_3$ followed by base catalyzed formation of $(Me_3Ge)_2$ Si $(SiC1_3)_2$ and related branched stannyl-silanes—Muller L., DuPont W., Ruthe F., Jones P.G. and Marsmann H.C. *J Organomet Chem.*, 579(1–2), 156–163 (1999).
276. Hydrosilylation and disilylation of cyclopropyl—substituted acetylene—Donskaya N.A., Lukovskii B.A., Mishechkin R.A., Yureva N.M. and Beltskaya I.P. *Russ. J. Org. Chem.*, 33(6), 898–899 (1997).
277. Deprotonation of benzyl-isothiocyanate: A simple route to silylated benzyl isothiocyanates—Bradsma L., Nedolya N.A. and Trofimov B.A. *Mendeleev Commun*, 6, 232–233 (1997).
278. Synthesis of mono- and bis-silylhydroxylamines—Wolfgang R. and Klingebiel U.Z. *Anorg. Allg. Chem.*, 624(5), 859–864 (1998).
279. O-halosilyl-N, N-bis (trimethylsilyl) hydroxylamines: Synthesis and structure-Wolfgramm R., Klingebiel U. and Noltemeyer M. ibid, 624(5), 865–871 (1998).

280. A steeply pyramidal silylamine (N_2O-dimethyl-N-silyl-hydroxylamine)—Mitzel N.W. and Oberhammer H. *Inorg. Chem.*, 37(14), 3593–3598 (1998).
281. Silyl-carbocyclization of 1,6-diyenes—Ojima I., Zhu J., Vidal E.S. and Kass D.F., *J.Am Chem. Soc.*, 120(27), 6690–6697 (1998).
282. Silylation reaction of 1,2-disubstituted alkenes catalyzed by a cationic zircocene complex (sequential cyclization)—Molander G.A. and Corrette C.P. *Tetrahedron Lett*, 39 (28), 5011–5014 (1998).
283. *Preparation of polymerizable triorganosilyl unsaturated carboxylates*—Fujino J., Taniguchi H., Morik K., Funaoka S. and Ito M. Jpn. Kokkai Tokkyo Koho JP10 195,084 (28 July, 1998).
284. Trimethylsilylation of natural silicates—Khananashvili L., Beroshvili M., Tsomaia N., Achelashvili V., Mtsvetadze L. and Chedia R. *Bull Georgian Akad Sci.*, 153(1), 52–54 (1996).
285. Silicon rid silylating agent: Synthesis of (N,N-dimethylamino) pentamethyl disilane—Hu C. and Guo Zh. *Huaxue Tongbao*, 1, 33–35 (1998).
286. Trichlorosilane/Triethylamine, an alternative to hexachlorodisilane in reductive trichlorosilylation reaction?—Muller I.P., Zamin A., Jeske J., Jones P.G. and DuMont W.W. ibid, 286–290 (1998).
287. Super silyl metal compounds—Wiberg N., Amelunxen K., Noth H., Apple A., Schmidt M. and Pullburn K. *Organosilicon Chem. III*, 3rd 1996 (Muench Silicon tage), pp 152–156 (Pub 1998).
288. Tetraanion tetra-lithium with 8 center/12 electron pi-system stabilized by silyl group: Synthesis and characterization—Sekiguchi A., Matsuo T. and Akaba R. *Bull Chem Soc Japan*, 71(1), 41–47 (1998).
289. Silyl carbonylation of 1, 5-dienes accompanied by acyl radical cyclization—Ryu I., Nagahara K., Komatsu M., Sonoda N. and Kurihara A. *J Organomet Chem.*, 548(1), 105–107 (1997).
290. Trimesityl silylium cation—Kresge A.J. *Chemtracts*, 10(11), 841–843 (1997).
291. Highly stable silyl radicals $(Et_nMe_{3-n}Si)_3$.Si (n = 1-3)—Kyushin S., Sakurai H., Betsuyaku T. and Matsumoto H. *Organometallics*, 16(25), 5386–5388 (1997).
292. Silylation of some aromatic polyamines—Cao S., Xu S., Zheng S. and Li Z. *Zheng-zhou Daxue Xuebao Ziram Kexueban*, 29(2), 992–94 (1997).
293. Unusual double silylation reaction of a $PtSi_2P_2$ Complex with an O-carbonyl unit—Kang Y., Kang S. and Ko J. *Organometallics*, 18(10), 1818–1820 (1999).
294. *Preparation of oligosilanyl-enolethers*—Yamamoto Y., Matsui T., Mori S. and Fakusima M. Jpn Kokkai Tokkyo Koho, JP11 147,889 (2 June, 1999).
295. Synthesis of aminomethyl-substituted silacyclohexanes from divinyl silane—Barfacker L., Elton D. and Eilbracht P. *Tetrahedron Lett*, 40(21), 4031–4034 (1999).
296. A stereospecific preparation of (E)-1, 1-dimethyl-2methyledene-silachromen by platinum catalyzed intermolecular hydrosilylation—Sashid H. and Kndoda A. *Synthesis*, 6, 921–923 (1999).
297. *Preparation of silanol containing organic silicon compound*—Ohkawa T. (Dow Corning Torray Silicone Co., Japan). Kokkai Tokkyo Koho, JP11 209, 382 (3 August, 1999).
298. *Preparation of gama-aminopropyl-silyl containing organosilicon compound*—Tachikawa M. ibid, JP11 209, 384 (3 August, 1999).
299. Reaction of bis (trimethylsilyl) sulfate with hexamethyldisilizane, acetamide and guamidine-hydrochloride—Belousova L.I., Vlasova N.N. and Voronkov N.G. *Russ. J. Gen. Chem.*, 68(3), 397–399 (1998).
300. Tris (trimethylsilyl) silane in organic synthesis—Chatgilialoglu C., Ferreri C. and Ginisis T. *Chem. Org. Silicon Compd.*, 2(Pt. II), 1539–1579 (1998).
301. Hydrosilylation of acetophenone with diphenylsilane in the presence of rhodium and Platinum complexes—Reznikov A., Lobadyuk V.I., Spevak V.N. and Skvortsov N.K. *Russ. J. Gen. Chem.*, 68(6), 910–913 (1998).

302. Trialkylsilyl-diazomethane derivatives: Wonderful chemical building block-Bertrand G. *Organosilicon Chem. III Muench Silicontage 3rd*, 1996, 223–236 (Published, 1998).
303. Reaction of hydridosilylamides—Junge K., Peulecke N., Sternberg K. and Reinke H. ibid, pp 353–357 (1998).
304. Preparation of chlorosilyl-enolates—Denmark S., Stavenger R.A., Winter S.B.D., Wong K.T. and Barsanti P.A. *J. Org. Chem.*, 63(25), 9517–9523 (1998).
305. Synthesis and reactivity of a stable Silylene—Haaf M., Schmiedl A., Schmedake T.A., Powell D.R., Millevolte A.J., Denk M. and West R. J. Am Chem. Soc., 120 (49), 12714–12719 (1998).
306. Thermally stable silylene (cyclic) Si $N(CH_2Bu)_2C_6H_4$-1,2: reactivity towards CN double bonds—Gehrhus B., Hitchcock P.B. and Lappert M.F. *Organometallics*, 17(7), 1378–1382 (1998).
307. Synthesis and properties of sila functional oligosiloxanes : a simple and practical method for the synthesis of 1,1,3,3-tetraisocyanato-1,3-disubstituted disiloxane—Abe Y., Abe K., Watnabe M. and Gunji T. *Chem. Lett.*, 3, 259–260 (1999).
308. Synthesis & structure of N-(silylalkyl) amides: rhodium catalyzed hydrosilylation of enamides—Murai T., Kimura F., Tsutsui K., Haisengawa K. and Kato S. *Organometallics*, 17 (5), 926–932 (1998).
309. Cyclic silyl-hydrazines and their borane adducts—Mitzel N.W., Hoffman M., Angermeir K., Schier K., Schlayer P. and Schmidbaur H. *Inorg. Chem.*, 34(19), 4840–4845 (1996).
310. *Preparation of tri (lower alkyl) silylethyl sulfides for formation of thin gold film on which organic thio groups are adsorbed*—Shimada S., Ohmsi S., Matsuda H., Nakanishi F. and Takeda H. Jpn. Kokkai Tokkyo Koho, JP11 100,387 (13 April 1999).
311. *Preparation of silyl-enolethers using nitriles solvents*—Yahata T., Endo M., Kubota T. and Tonumura Y. ibid, JP11 116,582 (27 April, 1999).
312. Silylium ion (R_3Si^+) problem: Bridging organic and inorganic chemistry—Reed C.A., *Acc. Chem. Res.*, 31(6), 325–332 (1998).
313. Synthesis and properties of dithinosiloles—Ohahita J., Nodono M., Watnabe T., Ueno Y., Kunai A., Harima Y., Yamashita K. and Ishikawa M. *J. Organomet Chem*, 553(1–2), 487–491 (1998).
314. Novel tetramerization of 1–trimethylsilyl–2-phenyl-cyclopropene—Lee G.A., and Chang C.Y. *Tetrahedron Lett*, 39(19), 3013–3016 (1998).
315. Synthesis and structure of N, N-bis (trimethylsilyl) trifluoromethane acidamide—Jonas S., Westerhausen M. and Simchen G. *J. Organomet. Chem.* 548(2), 131–197 (1997).
316. Dimerization of silaethylene—Venturini A., Bernadi F., Olivucci M., Rebb M.A. and Rossib I. *J. Am. Chem. Soc.*, 120(8), 1912–1913 (1998).
317. *Preparation of alkoxysilylpropionic acid esters*—Kozai T. and Kimura T. (Shin-Etsu Chemical Industries, Japan). Jpn. Kokkai Tokkyo Koho, JP10 17, 577 (20 January 1998).
318. Simplified preparation of trialkylsilyl and bis (trialkylsilyl) dihalo-methane via the deprotonation of dihalomethane—Yoon K. and Son D.Y. *J Organomet Chem*, 545–546, 185–189 (1997).
319. *Method for preparation of ketene-silyl-acetals*—Yanagase A., Tone S. and Tokimitsu S. Jpn. Kokkai Tokkyo Koho, JP10 17, 576 (20 January 1998).
320. Reactions of stable bis (amino) silylene (cyclic) $Si(NCH_2Bu)_2C_6H_4$-1,2 with multiply bonded compounds—Gehrus B. and Lapport M.F. Polyhedron, 17(5–6), 999–1000 (1998).
321. Synthesis of 1-functionalised silyl-cyclo-propanes via cyclization of 3-silylalk-4-enols—Nowak A. and Schumann E. *Synthesis*, 6, 899–904 (1998).
322. Hydrosilylation of unsaturated (hetero) aromatic aldehydes and related compounds catalyzed by transition metal complex—Iovel I., Popelis J. and Gangkhmann A. *J Organomet Chem.*, 559(1–2), 123–130 (1998).

323. Preparation of O-silylbenzyl alcohol-Hiji Y., Hudrlik E., Okuro C.O., and Hudrlik A.M. *Synth. Commun.*, 27(24), 4297–4308 (1997).
324. Trimethylsilyl silylium cation, verification of a free silylium cation in solution by NMR—Kranke E., Sosa C.P., Grafenstein J. and Cremer D. *Chem. Phys. Lett.*, 277(1,2), 9–16 (1997).
325. A supercharged silyl substituted 8 center anion, 12 electron pi-system-Sekuguchi A., Matsao T. and Kabuto C. *Angew Chem.* (Intl Ed), 36(22), 2462–2464 (1997).
326. A convenient synthesis of ethanobridged disilacyclo-octadiene—Haberhauer G. and Roers R. *Tetrahedron Lett.*, 38(50), 8679–8682 (1997).
327. Synthesis and reactivity of several stable 1-silaalenes—Trommer M., Miracle G.E., Eichter B.E., Powell D.R. and West R. *Organometallics*, 16(26), 5737–5747 (1997).
328. AMI treatment of silacyclacenes—Tuerker L. *Polycyclic Aromatic Compds*, 12(3), 213–219 (1997).
329. Synthesis and use of N-(trimethylsilyl)imines—Panunzio M. and Zarantonelo P. *Org Process Res & Dev*, 2(1), 49–59 (1998).
330. Hydrosylilation of ketones catalyzed by dimethyl-zircocene—Yun S.S., Yang Y.S. and Lee S. *Bull Korean Chem. Soc.*, 18(10), 1058–1060 (1997).
331. Reaction of 2,2,4,5-tetramethyl-1,3-dioxa-2-silacyclo-hexane with di-isobutyl boronate—Kaznetsov V.V. *Russ J Org Chem*, 33(11), 1674–1675 (1997).
332. *Preparation of N, N-bis (trimehylsilyl) allyl-amine by disproportionation of N-trimethylsilylallylamine*—Kubota T., Endo M. and Numanami K. Jpn. Kokkai Tokkyo Koho, JP10 218,883 (18 August, 1998).
333. Synthesis and reduction of Octa-silyl (4) radialene (a di-amine with 8 center 10 electron pi-system)—Matsuo T., Sakiguchi A., Ichinobe M., Ebata K. and Sakurai H. *Bull Chem Soc Japan*, 71(7), 1705–1711 (1998).
334. Tris (tert-butylamino) dimethylsilyl methylsilane and its precursors—Findeis B Schubart M., Gade L.H. and Corey J. *Inorg Synth*, 32, 136–140 (1998).
335. Ring opening of silacyclobutane—Gordon M.S., Barton T.J. and Nakano H. *J Am Chem Soc*, 119(49), 11966–11973 (1997).
336. Tris (trimethylsilyl) silamides of the heavier alkali-metals; a structural study—Klinkhammer K.W. *Chem Eur J*, 3(9), 1418–1431 (1997).
337. *Preparation of silaheterocycles, novel silaheterocycles, their application in production of silylene compounds as well as the resulting silylene derivatives*—Karsch H.H. and Schulueter P. Ger Offen DE19 711,154 (30 Oct, 1997).
338. Nucleophilic and electrophilic addition to silylated ketenimines generated from imidothioesters—Fromont C. and Masson S. *Phos Sulf, Silicon Rel. Elem*, 120–121, 397–398 (1997).
339. Using hydrosilylation to assemble organometallic polymers containing combination of silicon based functional groups—Kuhnen T., Ruffolo R., Stridiotto M., Ulbrich D., McGlinchey M.J. and Brook M.A. *Organometallics*, 16(23), 5042–5047 (1997).
340. Preparation of oligosilanes containing perhalogenated silyl groups ($-SiX_3$,$-SiX_2$, $-Six$ where X = C1, Br)—Herzog U. and Roewer G. *J. Organometallic Chem.*, 544(2), 217–223 (1997).
341. Synthesis, properties and structure of Poly (silyl)Pyridines: The phantom of intramolecular Si-N bonding—Redimiller F., Jockish A. and Scmidbauer H. *Organometallics*, 17(20), 4444–4453 (1998).
342. Bis (di-isopropylamino)silylene and its dimer—Tsutsui S., Sakamoto R. and Kira M. *J Am Chem. Soc*, 120(38), 9955–9956 (1988).
343. Relative rate of hydrosilylation of representative alkenes and alkynes by cp_2 YMe. THF—Molander G.A. and Knight E.E. *J Org Chem*, 63 (20), 7009-7012 (1998).
344. Lithium hydridosiloxy silylamides (reaction in n-octane and THF in presence of chlorotrimethyl-silane)—Junge H. and Popowski E. Z. *Anog Allg Chem.*, 623(9), 1475–1482 (1997).

345. Synthesis, structure and property of tricyclotetrasila-chalcogens—Unne M., Kawai Y. and Shiyoma H. *Organometallics*, 16(20), 4428–4434 (1997).
346. Generation and reactivities of phenylsilylene—Lee D.N., Kim C.W. and Lee M.E. *Phos Sulf Silicon Rel Elem*, 119, 37–47 (1996).
347. Gas phase generation and photoelectron spectrum of 1,1-dimethyl-N-dimethyl silysilanimine—Metail V., Joanteguy S., Chrostowska S.A., Pfister G.S., Systerman A. and Ripoll J.L. *Main Grp. Chem.*, 2(2), 97–106 (1997).
348. Intermolecular cordination in the solution of N-(dimethylhalosilyl methyl) amides and lactama—Negrebetaku V.V., Kramarova E.P., Shipov A.G. and Bankov.Y.I., *Russ J. Gen. Chem.*, 67(8), 1221–1232 (1997).
349. Selective deprotection of alkyl vs. aryl silyl ethers—Lipshutz B.H. and Keith *J. Tetrahedron Lett*, 39(17), 2495–2498 (1998).
350. *Gas phase reaction of silyl cation with water*—Leblac D., Cohen M.P. and Parker D.K. Eur Pat Appl EP 785–207 (23 July, 1997).
351. *Preparation of 3-chloropropy-trichlorosilane using hydrosilylation catalyst tertiary amines and colloidal platinum*—Takenchi M., Endo M., Kuboto T., Kiyomuri A. and Kuhoto Y., Jpn. Kokkai Tokkyo Koho, JP09 192, 494 (29 July, 1997).
352. *Phosphadisilacyclobutene, by stepwise silylene addition to phosphaalkynes*—Weidenbruch M., Olthoff S., Peters K. and VonScheneiring H. ibid, 15, 1433–1434 (1997).
353. Improved preparation of N-triflates and their use in silylating alcohols to silyl-ether—Olha G.A. and Klump D.A. *Synthesis*, 7, 744–746 (1997).
354. Reactions of acylpolysilanes with organolithium and silyl-lithium reagents—Oshita J. and Ishikawa M. *Main Grp. Chem. News*, 4(4), 16–23 (1997).
355. Preparation of silyl–enolether–Yamamoto Y., Matsui T. and Shimizu H. *Russ. J. Gen. Chem.*, 10 (87), 672 (1998).
356. Aluminumtrichloride catalysed intramolecular cyclization and allyl-silylation of diallylsilanes—Cho B.K., Choi G.M. Jin J. and Yoo B.R. *Organometallics*, 16(16), 3576–3578 (1997).
357. Reactions of silylketenes with carbanions—Akai S., Kitegaki S., Matsuda S., Tsuzuki Y., Naka T. and Kita Y. *Pharm. Bull*, 45(7), 1135–1139 (1997).
358. Hydrosilylation and related reactions of silicon compounds—Marciniec B. *Appl. Homogeneous Cat: Organometallic Compound*, 1, 487–506 (1996).
359. Gas phase association reaction of trimethylsilylium, $(CH_3)_3\ Si^+$, with organic bases—Stone J.A. *Mass Spectrom Rev.*, 16(1), 25–49 (1997).
360. Experimental & Theoretical studies of Silacycloheptatrienyl cation formation from phenylsilane—Jarak R. and Shin S. *J Am Chem Soc*, 119(27), 6376–6383 (1997).
361. Lewis acid catalyzed trans-allylsilylation of unactivated alkynes—Yoshikawa E., Gevorggyan V., Assao N. and Yamato Y. *J Am Chem*, 119(29), 6781–6786 (1997).
362. Nonionic superbase catalyzed silylation of alcohols—D'Sa B.A., McLeod D. and Verkade J.G. *J Org. Chem.* 62(15), 5057–5061 (1997).
363. Synthesis and reactions of silaboranes—Wesemann L., Ramjoie Y. and Trinkaus M. *Advances in Boron Chemistry* (Special Publication Royal Society of Chemistry) 201, 422–425 (1997).
364. *Preparation of cyclic-silylenolethers from silacyclobutanes and acid halides*—Yamashita M. and Tanaka M. Japan Kokkai Tokkyo Koho, JP09 136,894 (27 May 1997).
365. Electron transfer reaction of 1,2-disila-3,5-cyclohexadiene—Kako M, Takada H and Nakadaira Y. *Tetrahedron Lett.*, 38(20), 3525–3528 (1997).
366. *Preparation of (Chlorodialkyl silyl) (2 Chlorodialkyl silylethyl) benzenes*—Kabota T, Endo M. and Hirahara T. Japan Kokkai Tokkyo Koho, JP09 77,776 (25 March, 1997).
367. Forming new cyclotrisilanes and siliconium ions through reaction of cyclo-trisilane with Lewis acid—Belzner J, Ronneberger V, Schar D and Bonnecke C. *J. Organometallic Chem.*, 577(2) 330–336 (1999).

368. (8-methoxynaphthyl) silyl-triflates: A way to new functional siloxanes—Castel A, Rivierra P, Cosledam F and Gormitzka H. *C.R. Akad Sci. Ser IIc Chim*, 2(4), 221–228 (1999).
369. New method for introduction of a silyl group into alpha, beta-enones using a disilane catalyzed by a copper salt—Ito H, Ishizuka T, Tateiwa J., Sonada M and Hosomi A. *J Am Chem Soc.* 120(43), 11196–11197 (1998).
370. A new method for generation of dimethylsilanone, under mild condition—Voronkov M.G., Tsyrendorzhieve I. P, Ivanonva I.P and Dubenskaya E.I. *Russ J Gen Chem.*, 68(4), 658–659 (1998).
371. The first stable cyclotrisilene—Iwamoto T, Kabuto C and Kira M, *J Am Chem Soc*, 121(4), 886–887 (1999).
372. The first Si-H-B bridge (combination of 1,1-organoboration and hydrosilylation)—Wrackmeyer B, Tok O.L and Bubanov Y.N, *Angew chem* (International Edition), 38(1–2), 124–126 (1999).
373. The first allenic compound with doubly bonded P & Si: ArP:C:Si (Ph) Tip—Renaivonjatova H, Escudie J, Dubour A, and Declereq J.P. *Chem Eur J.*, 5(2), 774–781 (1999).
374. The S_Hi reaction at silicon: A new entry into cyclic alkoxysilanes —Studer A and Stean H. *Chem Eur J.*, 5(2), 759–773 (1999).
375. HCSiF and HCSiCl, the first detection of molecule with C ⋮ Si triple bond—Karni M, Apeloig Y, Schroder D, Zummack W, Rabezzana R and Schwarz H. *Angew Chem. (International Edition)*, 38(3), 332–335 (1999).
376. New Synthesis of silyl cyclopropanols via titanium mediated coupling of vinylsilanes and esters—Mizojiri R, Urabe H and Sato F. *Tetrahedron Lett.*, 40(13), 2557–2560 (1999).
377. New dicationic silicon complexes with N-methylimidazole—Hensen K, Kettner M, Pickel P and Bolte M. *Z Naturforsch (B), Chem Sci.*, 54(2), 200–208 (1999).
378. New method for synthesis of hexamethyl disilazane—Xu Zh, Wang M, Yu T, Kang H, Shi L and Zhao D. *Huaxue Shiji*, 21(3), 184–186 (1999).
379. High pressure synthesis of new silicon containing heteroatom analogs of fused norbornenes—Kirin S.I, Kalerner E.G and Eckert M.M *Synlett*, 3, 351–353 (1999).
380. A new class of potent protease inhibitors (silane-diol)—Sicburh S.M, Nittoli T, Mutahi A.M and Guo L. *Angew Chem* (International Edition), 37(6), 812–814 (1998).
381. First exclusive Endo-dig carbocyclization: $HfCl_4$ catalyzed intramolecular allylsilation of alkynes—Imamura K, Yoshokawa E and Gevorgyan V. *J Am Chem Soc.*, 120(21), 5339–5340 (1998).
382. Synthesis of new cyclolinear permethyloligosilane siloxanes—Chernyavskaya N.A, Alekksynskaya V.I and Chernyavski A.I. *Russ Chem Bull.*, 47(3) 526–527 (1998).
383. Synthesis of the first kinetically stable dibenzosilafulvene—Zemlynski N.N, Borisova I.V, Sheshatakova A.K, Ustynyuk Y.A and Chernysev E.A. *Russ Chem Bull.*, 74(3), 469–474(1998).
384. Iminosilanes as precursors of new rings and unknown ring system—Klinge-biel U and Niesmann J. *Phos. Sulf. Silicon Rel Elem*, 124–125, 113–122 (1997).
385. New cyclic stannyl, oligosilane—Kayser C, Klassen R, Schurmann M and Uhlig F. *J Organomet Chem.*, 556(1–2), 165–167 (1998).
386. Synthesis and catalysis of new organosilyl transition metal complex (activation of Si-Si sigma-bond by transition metal complexes)—Suginome M and Ito Y. *J Chem Soc, Dalton Transac.*, 12, 1925–1934 (1998).
387. New dichlorosilanes, cyclotrisilanes and silacyclopropane as precursors of intermolecularly coordinated silylenes–Belzner J, Dehnert U, Ihmels H, Huber M and Muller P. *Chem Eur J.*, 4(5), 852–863 (1998).
388. First experiment evidence for the formation of a silicate anion by intramolecular addition of a persulfoxide to a trimethylsiloxy group—Clannes E.L and Dillon D.L. *Tetrahedron Lett.*, 39(38), 6827–6830 (1998).

389. Synthesis of the first phthalocyanine containing dendrimer—Kraus G.A and Lousw S.V. *J Org Chem.*, 63(21), 7520–7521 (1998).
390. New chelating silylamido ligands: Synthesis and structure of lithium and magnesium derivative of t-BuHNSiMe$_2$-O-C$_6$H$_4$X (X = Ome, NMe$_2$, CH$_2$NMe$_2$, CF$_3$)—Goldfuss B, VonRague S, Handschuh S and Hampel F. *J Organomet Chem.*, 552(1–2), 285–292 (1998).
391. New trends of acylsilane chemistry at the service of organic chemistry—Bonini B.F, Franchim M.C and Fochi M. *Gazz Chim Ital.*, 127(10), 619–628 (1997).
392. A new and easy synthesis of bis (trimethylsilyl) ketone—Pan M and Bennecte T. *Synth Commn.*, 28(8), 1415–1419 (1998).
393. A new method for following the kinetics of the hydrolysis and condensation of silanes—Lindberg R, Sundholm G., Oya G and Sjoblom J. *Colloids Surf* (A), 135(1–3), 53–58 (1998).
394. New silahetrocycles: Formation and properties—Kroke E and Weidenbruch M. *Organosilicon Chem III, Muench Silicontage 3rd*, 1996, pp 95–100 (Pub. 1998).
395. A new route to silahetrocycles: Nucleopillic aminomethylation—Karsch H.H and Schreiber K.A; ibid, pp 237–240 (1998).
396. Controlled cleavage of R$_8$Si$_8$O$_{12}$ framework: a revolutionary new method of manufacturing precursors to hybrid inorganic-organic materials—Fehr F.J, Soulivong D and Eklund A.G. *Chem Commun (Cambridge)*, 3, 399–400 (1998).
397. A new and easy route to polysilanyl-potassium compound—Marschner C. *Eur J Inorg Chem.*, 2, 221–226 (1998).
398. Functionalised trisilylmethanes and trisilylsilanes as precursors of a new class of tripodal amido ligands—Schubart M, Findeiss B, Memnier H and Gade L. *Organosilicon Chem III (Muench Silicontage) 3rd*, 1996, pp 172–177 (1998).
399. First structurally defined catalyst for the asymmetric addition of trimethylsilyl-cyanide to benzaldehyde—Tararov B, Hibbs D.E and Hursthouse M.B. *Chem. Commun (Cambridge)*, 3, 387–388 (1998).
400. Methoxy (bis) (tris) (trimethylsilyl) methane: The first germinal di (hypersilyl) compound—Jeschke E, Gross T, Reicke H and Ochen H. *Organosilicon Chem III (Muench Silicontage) 3rd*, 1996, pp 178–181 (Pub 1998).
401. A new example of linear disiloxane: Synthesis and X-ray crystal structure of bis (2-silolyl) tetramethyl-disiloxane—Yamaguchi S, Jin R. Shoro M and Tamao K. *Chem Heterocycl Compd.*, 33(2), 155–160 (1997).
402. A novel organo-lead(II) compound with silicon—Eaborn C, Ganicz T, Hitch- cock P.B, Smith J.D and Soezerli S.E. *Organometallics*, 16(26), 5621–5622 (1997).
403. Magnetic organometallosiloxane—Leviskii M.M and Buchaecenko A.L. *Russ Chem Bull.*, 46(8), 1367–1378 (1997).
404. Novel water soluble organosilane compound as a radical reducing agent in aqueous media—Yamazaki O, Togo H, Nogami G and Yokoyama M. *Bull Chem Soc Jpn.*, 70(10), 2519–2523 (1997).
405. A new spirocyclic system: Synthesis of a silaspirotropylidene—Sohn H, Merritt J, Powell D.R and West R. *Organometallics*, 16(24), 5133–5134 (1997).
406. First synthesis and characterization of stable heteroleptic silylstannylenes—Drost C, Gehrhus B, Hitchcock P.B and Lappert M.F. *Chem Commun (Cambridge)*, 19, 1845–1846. (1997).
407. *Preparation of new silane compounds as coupling agents for inorganic particles*—Kawai O and Nekanchi J. Jpn Kokkai Tokkyo Koho JP09 255,690 (30 Sept., 1997).
408. New chemistry of alpha-silylvinylsulfides—Bonini B.F, Comes F. M and Fochi M. *Phos Sulf, Silicon Rel Elem*, 120–121, 451–452 (1997).
409. Silicate cage: Precursors to new materials—Harrison P.G. *J Organomet Chem.*, 542(2), 141–183 (1997).

410. Synthesis of new functionalised bisacylsilanes—Boullion J.P and Portella C. *Tetrahedron Lett.*, 38(37), 6595–6598 (1997).
411. Silacyclopentadienylidene: The first silylene incorporated in a silole ring—Kako M, Oba S, Uesugi R, Sumiishi S, Nakadaira Y and Tanaka T. *J Chem Soc (Perkin Transac II)*, 7, 1251–1253 (1997).
412. The first silatrane with a direct Si-NCS bond: 1-socyanato-silatrane—Narula S.P, Shanker R, Kumar M, Chaddha R and Jainak C. *Inorg Chem*, 36(17), 3800 (1997).
413. A new synthetic route to allylsilanes—Saito S, Shimada K, Yamamato H, Marigorda E.M and Fleming I. *Chem Commun (Cambridge)*, 19, 1299–1300 (1997).
414. Photochemical functionalisation of C_{60} with phenylpolysilane—Kusukawa T and Ando W. *J Organomet Chem*, 559(1–2)11–22 (1998).
415. *Preparation of silyl groups containing fullerenes with improved optical function*—Yoshida T and Mori S (Nippon Kayaku Co, Japan). Jpn Kokkai Tokkyo Koho JP09 157,273 (17 June, 1997).
416. Hydrosilylation of fullerene (C_{60})—Bespalova N.B, Bivina M.A, Rebrov A. I, Khodzaeva V.L and Semenov O.B. *Russ Chem Bull*, 46(9), 1620–1621 (1997).
417. Novel Fullerene based organosilicon compounds—Akasaka T, Kabayashi K, and Nagasa S. *Proc Electrochem Soc* 97–42(5), 283–288 (1997).
418. Reactions of fullerenols, C_{60} $(OH)_x$ (X = 12,18), with trialkoxysilane, $(RO)_3Si$ $(CH_2)_3X$ (R = Me, X = Cl; R = Et and X = NH_2)—Iglesias M and Santos A, *J Organomet Chem*. 553 (1–2), 193–197 (1998).
419. Chemical derivitization of fullerene with organosilicon compounds—Akasaka T, Kobayashi K, Nagase S and Suzuki T. *Proc Electrochem Soc*. 97–114, 265–270 (1997).
420. Organosilicon (polysilane) derivatives of fullerene—Ando W and Kasakawa T. *Chem Org Silico Compd* 2 (Part III), 1929–1960 pp (1998). Wiley Chichester U.K (Ed by Rapport Z and Apeloig Y).
421. Synthesis and structure of silicon-doped heterofullerenes—Ray C, Lerme J, Pellarin M, Vialle J.L, Broyer M, Blase X, Melinon P, Keghelian P and Perez A. *Phys Rev Lett.*, 80(24), 5365–5368 (1998).
422. Substituent effect on addition of silyl-lithium and germyl-lithium to fullerene (C_{60})—Kusukawa T and Ando W. *J Organomet Chem.*, 561(1–2), 109–120 (1998).
423. The exohedral modification of fullerene with several silicon containing adducts and synthesis as well as characterization of sila-fullerenes—Miller M.L; *Diss Abstr Intl* (B), 59(2), 668 (1998).
424. Organosilicon *compounds and their use in forming homotropically oriented liquid crystalline phases on surfaces*—Panluth D, Bremer M, Bochm E., Osabe A, Herget G and Hechler W. Ger Offen DE19, 848,800 (20 May, 1999).
425. *Preparation of silacyclohexanones and their intermediate silanes as an intermediate for liquid crystals*—Takeda T, Kitahara T, Watnabe H, Kaneo T, Hagiware T and Shimizu T. Jpn Kokkai Tokkyo Koho JP10 195,079 (28 July 1998).
426. *Preparation of silacyclohexanones* as intermediates for liquid crystals—Kitahara T, Takeda T, Ogiwara T and Kiyomi T (Shin-Etsu Chemical Industries); Jpn Kokkai Tokkyo Koho, JP09 157,274 (17 June, 1997).
427. *Preparation of silacyclohexanones as intermediate for liquid crystals*—Takahashi T, Asakura K, JP09 194,486 (29 July, 1997).
428. *Preparation of (Fluoroalkoxy) alkoxy-silane as material for SiOF interlayer insulator film for semiconductor devices by CVD*—Hijido T, Kadokura H , Matsumoto H and Yokoyama H. ibid, JP10 101,682 (21 April, 1998).
429. Photoinduced intramolecular cyclization of 1-(O-allyloxyphenyl)-2-pentamethyl-disilanyl-ethyne—Shim S.C. and S.K. *Tetrahedron Lett.*, 39(38), 6891–6894 (1998).
430. Photoreaction of 1-(O-acetoxy phenyl)-2-pentamethyl-disilanyl-ethyne—Shim S.C.and Park S.K. *Bull Korean Chem Soc*. 20(5), 547–500 (1999).

431. Preparation *of tetrathienylsilane derivatives for electrical, optical, or magnetic devices*—Nakayuma S, Hayashi K, Kurodo M and Tsutsui A. Jpn Kokkai Tokkyo Koho JP09 241, 265 (16 September, 1997).
432. *Manufacture of optically active halohydrin-trialkyl-silyl-ether*—Nugent W.A. PCT Intl Appl WO 99 02, 535 (21 January, 1999).
433. *Preparation of photofunctional hydrolyzable organosilicon compounds*—Yamada W, Nukada K, Ishii M. Jpn Kokkai Tokkyo Koho JP11 171, 890 (29 June 1999).
434. Photochemical sulfochlorination of tetramethyl-silane and hexamethyl-disiloxane—Bolshovkova S.A, Vlasova N.N, Pozidaev Y.N and Voronkov M.G. *Russ J Gen Chem*, 67(5), 712–714 (1997).
435. Photoinduced electron transfer reaction of 7,8-disilabicyclo (2,2,2) octa-2,5-diene—Kako M, Hatakenaka K, Kakuma S, Ninomiya M, Nakodaira Y, Yasui M, Iwasaki F, Wakasa M and Hayashi H. *Tetrahedron Lett.*, 40(6), 1133–1136 (1999).
436. Photoinduced silylene transfer reaction of cyclic organo-silanes to phenenthroquinone—Kako M, Ninomiya M and Nakadaira Y, *Chem Commun (Cambridge)*, 15, [illegible]3–1374 (1997).
437. The photochemistry of dimethylsilylazide—Kuhn A and Sander W. *Organometallics*, 17(2), 248–254 (1998).
438. Photochemical reaction of (pi-MeC_5H_5)$Mn(CO)_2L$ with $HMe_2SiXiMe_2H$—Schubert U and Grubert S. *Monatsh Chem.*, 129(5), 437–443 (1998).
439. Photoreaction dynamics of permethylcyclohexasilane $(Me_2SI)_6$ Via triplet state surface—Tachikawa H. *J Organometallic Chem*, 555(2), 161–166 (1998).
440. Photochemical reaction of silylene with ethene and silene—Lennartz C, Hildebrandt H.M and Engels B. *J Phys Chem A*, 101(51), 10053–10062 (1999).
441. A transient spectroscopic study of photochemistry of vinyldisilanes—Leigh W.J, Bradaric C.J, Slugett G.W, Venneri P, Coulin R.T, Durjati MSK and Ezhova M.B. *J Organomet Chem*, 56(1–2), 19–27 (1998).
442. The photo-physics of permethylated oligosilane chains—Raymond M.K. *Disser Abstr Intl B*, 58(7), 3642 (1998).
443. Photochemistry of organosilicon compounds—Brook A.G. *Chem Org Silicon Compd 2* (Part II), pp 1233–1310 (1998).
444. Mechanistic aspect of photochemistry of organosilicon compounds—Kira M, ibid, 1311–1337 (1998).
445. Preparation and photodecomposition of allyl and vinyl containing di- and oligo-silanes—Semenov V.V, Ladilina E.Y, Cherepnnikova N.F, Lopatin M.A, Khorsev S.Y and Kurskii Y.A. *Russ J Gen Chem*, 67(9), 1455–1462 (1997).
446. Electrochemical formation of cyclosilane—Graschy S, Grogger C and Hengge E. *Organosilicon Chem III (Muench Silicontage) 3rd*, 1996, pp 317–321 (Published 1998).
447. Synthesis as well as optical, electrochemical and electronic properties of silicon compound: bridged bithiophenes—Oshita J, Nodono M, Kai H, Watnabe T, Kunai A, Komguchi K, Shiotani M, Adachi A, Okita K, Horima Y, Yamashita K and Ishikawa M. *Organometallics*, 18(8), 1453–1459 (1999).
448. Electrochemical synthesis of Fluoro-organo-silanes—Martinov B.I and Stepnov A.A *J Fluorine Chem.*, 85(2), 127–128 (1997).
449. Electrosynthesis of alpha-silylalkylnitrile—Constatieux T and Picard J. P, *Synth Commun.*, 27(22), 3895–3907 (1997).
450. Electrochemistry of Tetramesityl-disilene, $Mes_2SiSiMes_2$—Zhang Z.R, and Becker J.Y. *Chem Commun.*, 24, 2719–2720 (1998).
451. Electrochemical oxidation of Benzylsilanes—Zhuikov V.V. *Russ J Gen Chem.*, 67(6), 887–893 (1997).
452. Electrosynthesis of new functionalised difluoroallylsilanes—Rajaonah M, Rock M.H, Bergne J.P, Bonett D.D, Condon S and Sylvic N.J. *Tetrahedron Lett*, 39(20), 3137–3140 (1998).

453. Electrochemistry of organosilicon compounds: From halosilanes to ethynyl-silane, siloles, and organosilicon polymers—Kunai A, Ohnishi O and Morishita M, *Novel Trends Electro-org Synth, Proc Intl Symp 3rd*, 1997, pp 363–366 (Published 1998).
454. Electroreductive synthesis of sequence ordered polysilanes using Mg-electrodes—Ishifune M, Bu H, Kashimura S, Yamashita N and Shono T. ibid, pp. 371–372 (Pub 1998).
455. Dissociative excitation of gaseous organosilicon compounds by electron impact—Morita H, Sasagawa H, Toh H and Sumida K. *J Photopolym Sci & Technol.* 11(1), 81–84 (1998).
456. *Preparation of siloxanes as bactericides and algicides*—Nagase H, Akimoto M, Aoyagi T, Tanaka K, Sano Y and Morita M (Sagami Chemical Research Center) Japan. Jpn. Kokkai Tokkyo Koho JP09 143,186(3 June 1997).
457. Synthesis and fungicidal activity of 1-(H-1,2,4-triazol-1-yl)alkyl-1-silacyclohexanes—Yoo B.R, Suk M.Y, Young M and Soon G. *Bull Korean Chem Soc*, 19(3), 358–362 (1998).
458. Synthesis and fungicidal activity of S-allyl-O-substituted phenylthio-phosphates (-onates)—He Z, Liu J, Zhu Z and Tang C. *Gaodeng Xuexiao Huaxue Xuebao*, 20(2), 227–231 (1999).
459. Synthesis and insecticidal activity of silicon containing asymmetric thiophosphates and their carbon analog—He Z, LiZ, Wang Y and Quian B. *Yingyong Huaxue*, 15(6), 29–32 (1998).
460. *Preparation of Difluorovinylsilane as insecticide and acaricide*—Barnes K.D and Fu Y.L. Jpn Kokkai Tokkyo Koho JP09 249,674 (22 September 1997).
461. An alternative synthesis of the potent anti-muscarinic agent: Silabiperiden—Pikies J and Ernst L. *Phos., Sulf., Silicon Rel., Elem.*, 128, 179–190 (1990).
462. *Silanyltriazines as light screening compound*—Huber U. Eur Pat Appl. EP 933,376 (4 August, 1999).
463. *Method of inhibiting vascular cell adhesion and treating chronic inflammatory disease with 2, 6-dialkyl-4-silylphenols*—Wright P.S and Busch S.J. US U.S 5, 795,876 (18 August, 1998).
464. *Silyl modification of biologically active compounds*—Lukevies S, Germane S, Segal I and Zablotskaya A. ibid, 33(2), 234–238 (1997).
465. Biocatalytic kinetic resolution of hydroperoxyvinylsilanes by horeseradish-peroxidase and lipases—Adam W, Mock K, Saha M and Chantu R. *Tetrahedron: Asymmetry*, 8(12), 1947–1950 (1997).
466. Synthesis and properties of novel silanediol-protease enzyme inhibitors—Chem C.A; *Disser Abstr Intl (B)*, 59(1), 224 (1998).
467. Process for preparing propyltrichlorosilane—Marciniec B, Kitynski D, Oczkowich S, Tyrka M, Lewandowski M and Foltynowich Z. *Przem Chem*, 77(9), 336–339 (1998).
468. *Improved method for obtaining Organosilanes using Lewis acid ($Ticl_4$, $FeCl_3$,CuCl, AgCl, $ZnCl_2$, $AlCl_3$, $SnCl_2$, $BiCl_3$) catalyzed redistribution reaction and an alkoxysilane or silicone-resin as a redistribution reaction inhibitor*—Cardinand D and Colin P (Rhodia Chimie, France). Fr Demande FR 2,761,360 (2 Oct, 1998).
469. A new route to silicon-alkoxides from silica—Kemmit T and Handerson W. *Aust. J Chem.*, 51(11), 1031–1035 (1998).
470. *Stabilization of purified Trimethoxysilane by storing in metal container*—Onosawa K and Watnabe T. Jpn Kokkai Tokkyo Koho JP11 124, 386 (11 May, 1999).
471. *Process for converting polymeric silicon containing compounds to monosilanes*—Crun B.R, Freeburne S.K and Wood L.H. U.S US 5,907,050 (25 May, 1999).
472. *Continuous process for preparing aminopropyl trialkoxysilanes*—Balduf T, Wieland S and Lortz W. Eur Pat Appl EP 849, 271 (24 June, 1998).
473. *Preparation of alkoxysilyl containing silatranes as coupling agent or adhesion promoter*—Yoshitake M, Hatakenaka H and Fukutani Y. Jpn Kokkai Tokkyo Koho JP10 182,669 (7 July, 1998).

474. Silsesquioxanes: A key intermediate in the building of molecular composite materials—Provatas A, Luft M, Mu J.C, White A.H and Matisona JG. *J Organomet Chem.*, 565(1–2), 159–164 (1998).
475. *Fluoroalkyl containing organosilicon compounds and their use*—Jenkner P, Frings A.J, Hoen M, Monkiewich J and Standke B. Eur Pat Appl Ep 838,467 (29 April, 1998).
476. *Process for the production of Octaphenylcyclotetrasiloxane*—Razzano J. S (General Electric). U.S. Pat 5,739,370 (14 April, 1998).
477. *Preparation of Trimethylchlorosilane*—Matveev L, Efimov Y.T, Maksimova G.V, Stepanova A.N, Simakov V.I, Razmakov E and Zheltukhin I.A. Russ RU 2,099, 343 (20 December, 1997).
478. *Cyclic siloxanes and their use as wetting aids and foam stabilizer*—Hanbennestel K and Bubat A. Ger Offen DE 19,631,227 (23 April, 1998).
479. *Organosilicon chemistry (from molecule to material)*—Norbert A and Weiss J (Ed), Wiley-VCH Pub, Germany (1998).
480. *Method for preparation of organosilanes using redistribution reaction catalyzed by Al_2O_3 based material*—Colin P. PCT Intl Appl WO98 43,984 (8 October, 1998).
481. *Preparation of organosilanes by redistribution reaction catalyzed by Lewis acids using catalyst inhibitor after redistribution*—Coordinand D. PCT Intl Appl WO98 43,985 (8 October, 1998).
482. *Polyfunctional organosilane coated silica*—Menon V.C, Wallace S, Maskara A, Smith D.M and Koehler K.C. PCT Intl Appl WO99 36,356 (22 July, 1999).
483. Managing a technical revolution: The switch from trichlorosilane to trimethoxysilane based processes—Ritscher J.S. *Silicon Chem Ind IV Conf Proc.*, pp 265–273 (1998).
484. Continuous organomagnesium synthesis of phenylethoxysilanes from a mixture of phenyltriethoxysilane with phenyltrichlorosilane—Klokov B.A (Joint Stock Co 'Silane', Lipetsk, Russia). *Main Group Metal Chem*, 22(1), 1–4 (1999).
485. The direct method to Methylchlorosilanes: reflections in chemistry and process technology—Pachly B and Weiss J. *Organosilicon Chem III 3rd conf 1996 Proc.* pp 478–483 (1998).
486. Synthesis and application of omega-epoxy-functionalized alkoxysilanes—Sperveslage G, Stoppeck L.K and Grobe J. ibid, 515–519 (published 1998).
487. Thermal stability (to –75°C) of 1-lithio-2-trimethylsiloxyethylene in diethylether solution—Bandrillard V, Ple G, Davoust D and Dunhamel L. *J Chem Res Synop.*, 12, 456–457 (1997).
488. *Procedure for production of Acyloxyalkoxysilanes*—Friedrich H, Lentener B, Mrongue N and Schmidt R (BASF-Germany). Ger Offen DE 19,632,483 (19 Feb, 1998).
489. Preparation and synthetic utility of Oxasilacyclopentane-acetals derived from siliranes—Shaw J.T and Woerpel K.A. *Tetrahedron*, 53(48), 16597–16606 (1997).
490. *Preparation of organosilicon compounds as surface treatment agents and resin additive*—Tsuchida K (Japan Energy K.K). Jpn Kokkai Tokkyo Koho JP09 295,989 (18 November, 1997).
491. Preparation of organosilicon compounds as surface treatment agent and resin additive—Tsuchida K (Japan Energy K.K) ibid, 09 295,991 (18 November, 1997).
492. *Process for preparation of alkynylsilanes by catalytic coupling reactions*—Fuchigami T and Shimizu R. ibid, 09 295,986 (18 November, 1997).
493. *High purity branched Phenylsiloxane fluids*—Legrow G.E and Lantham I.W. US U.S 5,679 822 (21 October, 1997).
494. *Preparation of organosilicon compounds*—Cheryshev E.A, Belyakova Z.V, Pomerantseva M.G. and Yagodine L.A. USSR SU 1,524,451 (10 February, 1997).
495. Synthesis of N-(trialkylsilyl) morpholins and their use as hydrophobic layer on silica—Fadeev A.Y and Mingalev P.G. *Vestn Moskow Univ. Ser II Khim*, 37(6), 588–591 (1996).

496. Cyclo-organosilyl derivative for determination of alcohols and carboxylic acids by G.C/Mass-spectra—Zaikin V.G, Yakushin V.N, Volmina E.A and Mikaya A.I. *Eur Mass Spectron.*, 5(1), 23–30 (1999).
497. Preparation of Allyltrialkoxysilanes as resin modifiers or coupling agent—Kuboto T, Endo M and Numanami K. ibid, 09 157,280 (17 June, 1997).
498. *Preparation of acetylene group containing silanes as materials for heat resistant polysilanes*—Ishikawa J, Inone K, Iwata K and Ito M (Mitsui Toatsu Chem. Japan). Jpn Kokkai Tokkyo Koho JP09 104,688 (22 April, 1997).
499. *Preparation of 1,4-dioxanyl group containing dihydroxysilane as an intermediate for hydrophillic silicones*—Shinohara N, Igarashi M and Takahashi M. ibid, 09 104,689 (22 April, 1997).
500. *Preparation of 1,4-dioxanyl group containing dialkoxysilanes as an intermediate for hydrophillic silicones*—Shinohara N, Igarashi M, Takahashi M, Hirai T, Banno T and Umeno M. ibid, 09, 104,690 (22 April, 1997).
501. *Purificaiton of alkoxysilanes*—Nishida M and Sugito T (Toshiba Silicone Co. Japan). ibid 09 151,190 (10 June, 1997).
502. *Process for purification of tetraethoxysilane using a chromatographic separation column*—Potts T. PCT Intl Appl WO 97 18,219 (22 May, 1997).
503. *Purification of Dimethoxymethylsilane by azeotropic removal of methanol*—Asai Y and Ogawa N (Kanegafuchi Chemical Industries, Japan). Jpn Kokkai Tokkyo Koho, JP09 77,778 (25 March, 1997).
504. *Purification of Dimethoxymethylsilane by removal of methanol with methylformate*—Asai Y and Ogawa N. ibid 09 77,779 (25 March, 1997).
505. *Purification of organosilanes of group 13 (IIIA) and group 15 (VA) impurities by complexation*—Laxman R.K (Air Products & Chemicals, USA). Eur Pat Appl EP 879,821 (1998).
506. *Preparation of siliconperoxide compounds*—Koenigstein K. Ger Offen DE 19 540,581 (7 May, 1997).
507. Model compound for Diphenylsiloxane polymer (Octaphenyltetra-siloxane-1,7-diol and its organo-tin derivative)—Beckman J, Jurkschat K, Mueller D and Rabes Schuerman M. *Organometalics*, 18(12), 2326–2330 (1999).
508. *Method for synthesis of chlorosilanes*—Wheeler D.R and Pollagi T.P. U.S. Patent 5, 939,577 (17 August, 1999).
509. *Preparation of alkoxysilanes from silicon and alcohols*—Hasagawa M *et al.* Jpn Kokkai Tokkyo Koho JP10 182,661 (7 July, 1998).
510. *Preparation of alkoxysilanes from silicon and alcohols*—Takazawa A and Sugaro Y (Mitsubishi Chemical Industries, Japan). ibid 09 87,287 (31 March, 1997).
511. *Preparation of alkoxysilanes with recycling alcohols*—Suguro Y, Utsunomaya A and Yasuda M. ibid 09 95,491 (8 April, 1997).
512. One vessel synthesis of N-(methyldihalosilylmethyloamides and lactam—Kramarova E.P, Shipov A.G, Negrebekski V.V and Bankov Y.I. *Russ J Gen Chem.* 67(8), 1315–1316 (1997).
513. *One step preparation of alkynylsilanes from halosilanes and 1-alkynes*—Kondo S, Sugita H, Hiyama T and Hatanaka Y. Jpn Kokkai Tokkyo Koho JP09 194,484 (29 July, 1997).
514. *One step preparation of 1-alkynylsilanes from halosilanes and alkynyl-copper*—Kondo S, Sugita H, Hiyama T and Hatanaka Y. ibid, 09 194,485 (29 July, 1997).
515. *Use of surface active additives in the direct synthesis of trialkoxysilanes*—Medicino F.D, Childress T.E, Magri S and Lewis K.M. Eur Pat Appl EP 835,877 (15 April, 1998).
516. *Preparation of organic silicone compounds containing sulfonic-acid groups as modifier and coupling agent*—Ichinohe S, Yamagisawa M and Takahashi M. Jpn Kokkai Tokkyo Koho, JP 09 165,303 (24 June, 1997).
517. *Preparation of organic silicon compounds as coupling agents*—Hayashi M and Iwai R (Dow Corning Torray Silicon Co, Japan). ibid, 09 136,893 (27 May, 1997).

518. *One step process for converting high boiling residue from direct process to monosilanes*—Freeburne S.K and Jarvis R.F. U.S US 5, 627,298 (6 May, 1997).
519. First step in the direct synthesis of methylchlorosilane: Transfer of metallic copper to silicon—Glasston T, Bertolini J.C and Colin P. *Silicon Indus IV Conf Proc* ed by Oeye H.A, pp 145–156 (1998).
520. *Process for preparing nitrogen or sulfur containing organosilanes*—Simonian A.K, Webb J.L, Brunelle D.S, Banach T.E and Rubinsztajn S. Eur Pat Appl EP 900,801 (10 March, 1999).
521. Continuous organomagnesium synthesis from mixtures of ethylethoxysilanes with methyl (thienyl) or haloorganoethoxysilanes—Klokov B.A. *Zh Prikl Khim*, 71(3), 461–464 (1998).
522. *Preparation of alkoxysilanes containing little quantity of halogens*—Ono T, Doi K, Yasutake T, Yoshino T and Sumi T, Jpn Kokkai Tokkyo Koho, JP10 287,682 (27 October, 1998).
523. Formation and molecular structure of $[(CH_3)_3Si_3C(C_3H_7)In(\mu\text{-}OH)]_3$—Walz A, Niemeyer M, Weidlein J. Z *Anog Allg Chemi*, 625(4), 547–549 (1999).
524. Formation of sila-zircon acyclopentene via zirconium-silene complex and alkyne—Mori M, Kuroda S and Dekura F. *J Am Chem Soc*, 121(23), 5591–5592 (1999).
525. Recent advances in the chemistry of silicon-heteroatom multiple bond—Tokitoh N and Okazaki R. *Chem Org Silicon Compd 2 (Parts II)*, 1063–1103 (1998).
526. Matrix isolation studies of silicon compounds (silicon heteroatom multiply bonded compounds)—Maier G, Mendt A, Jung J and Pacl H. ibid, 2 (Pt II), 1143–1185 (1998).
527. Hydroboration of vinylsilane—Parks D. J and Piers W.E. *Tetrahedron*, 54(51), 15469–15488 (1998).
528. *Preparation of Fluorosilanes*—Uchimara Y and Tanaka M. Jpn Kokkai Tokkyo Koho JP11 12,287 (19 January, 1999).
529. Five and Six membered Si-C hetrocycle: Synthetic method for construction of silacycle—Hermanas J and Schmidt B. *J Chem Soc (Perkin Transac)*, 14, 2209–2230 (1998).
530. Hetro pi-system: Silylenes of the elemental composition C_4H_2Si (generation and matrix spectroscopic identification)—Maier G, Resisenauer H.P and Mendt A. *Eur J Org Chem.*, 7, 1285–1290 (1998).
531. Hetro pi-system: Reaction of silicon atoms with HCN (production and matrix spectroscoric identification of CHNSi & CNSi isomers)—Mair G, Reisenauer H. P, Egenolf H and Glathaar J. ibid, 7, 1307–1311 (1998).
532. Reaction of silver (I) trifluromethanethiolate with halotrimethylsilane: *in-situ* generation of trimethylsilyltrifluoromethylsulfide—Adam D.J, Tavener S.J and Clark J.H. *J Fluorine Chem*, 90(1), 87091 (1998).
533. Reactions of lithium-hydridosilylamides with carbonyl compounds and mixtures of CO-compound and chlorotrimethylsilane—Scheneider J, Popowski E and Fuhrmann H. Z *Naturforsch(B). Chem Sci*, 53(7), 663–672 (1998.)
534. Synthesis of difunctional borasiloxanes—Mingotand A.F, Heroguez V and Soum A. *J Organomet Chem*, 560(1–2), 109–115 (1998).
535. Hetero pi-system: Silylenes of the elemental composition $C_2H_4Si_2$—Mair G, Reisenauer H.P and Mendt A. *Eur J Org Chem*, 7, 1291–1296 (1998).
536. Hetero pi-system: C_3H_2Si species—Mair G, Reisenauer H.P, Jung and Egenlolf H. ibid, 7, 1297–1305 (1998).
537. Novel silaphosphahetrocycles and hypervalent silicon compounds with phosphorus donors—Kersch H.H, Richter R and Witt E. ibid, pp 460–465 (1998).
538. Metastable compounds containing silicon-phosphorus and silicon-arsenic multiple bond: Synthesis, structure and reactivity—Driess M. Rell S, Winkler U and Pritzkow H. ibid, pp 126–143 (1998).
539. Novel pathways to the reaction of vinylsilanes with lithium metal—Maercker A, Reider K and Girreser U. ibid, pp 195–205 (1998).

540. Silyl group transfer from allylsilane to olefins catalysed by a rhuthenium complex—Kalkiuchi F, Yamada A, Chatani A, Murai S, Furukawa N and Seki Y. *Organometallics*, 18(10), 2033–2036 (1999).
541. Modification of the base catalyzed disproportionation of methylchlorodisilanes—Herzog U, Schulze N, Trommer K and Roewer G. *Main Grp Met Chem*, 22(1), 19–33 (1999).
542. Hydroformylation of vinylsilanes with rhodium (aceac)-CO_2/tris(N-Pyrolyl) Phosphine catalytic system—Trzecian A.M, Ziolkowski J.J and Marcience B. *C.R. Acad Sci Ser IIc: Chim*, 2 (4), 235–239 (1999).
543. Aminosubstituted disilanes: Synthesis by unsymmetrical reductive coupling Heinicke J and Mantey S. *Hetroatom Chem*, 9(3), 311–316 (1998).
544. Hexachlorodisilane initiated gas phase reaction of tetrachlorogermane with ethyl and butylchlorides—Chernyshev E.A, Komalenkova N.G, Yakovlev G.N, By-Kovchenko V.G, Khromykh N.M and Bochkarev V.M. *Russ J Gen Chem*, 68(3), 403–406 (1998).
545. Carbenoid insertions into the silicon-hydrogen bond catalyzed by copper Schiff-base complexes—Dakin L.A, Schaus S.E, Jacobsen E.N and Panek J.S. *Tetrahedron Lett*, 39(49), 8947–8950 (1999).
546. Stereoselective synthesis of silylenolethers via the irridium catalyzed isomerisation of allyl-silyl-ethers—Ohmura T, Yamamoto Y and Miyaura N. *Organometallics*, 18(3), 413–416 (1999).
547. Cyclopropanation of the enoxysilane under ultrasound irradiation—Sen H.Z and Li J.S. *Youji Huaxue*, 18(6), 550–555 (1998).
548. Silicon effect favouring the formation of a cyclopentene via palladium catalyzed 5-endo-trig cyclization—Thorimbert S and Malacria M. *Tetrahedron Lett* 39(52), 9659–9660 (1998).
549. Titanocene catalyzed double silylation of dienes and aryl-alkenes with chlorosilanes—Terao J, Kambe N and Sonoda N. *Tetrahedron Lett*, 39(52), 9697–9698 (1998).
550. Metastable decomposition of $(CH_3)_3SiOCH_3$, $(CH_3)_3SiSCH_3$ $(CH_3)_3SiCH_2OH$ $(CH_3)_3SiCH_2SH$ upon electron impact—Tajima S, Sekiguchi O, Kinoshita T and Shigihara A. *Adv Mass Spetrose* 14, 13050–13051 (1998).
551. Stereospecific and regioselective reaction of silyacyclopropane with carbonyl compounds catalyzed by copper-salts (evidence of a transmutation mechanism—Franz A.K and Woerpel K.A. *J Am Chem Soc*, 121(5), 949–957 (1999).
552. *Preparation of polycyclic aminodialkoxysilanes as polymerization catalyst*—Ikai S, Fukumaya T and Sakagami S. Jpn Kokkai Tokkyo Koho JP11 106,390 (20 April, 1999).
553. Catalysis of siliconalkoxide transesterification by early transition metal complexes—Curan M.D, Gedria T.E and Stiegman A.E. *Chem Mater*, 10(6), 1604–1612 (1998).
554. Solid acid catalyzed disproportionation and alkylation of alkylsilanes—Yamaguchi T, Shibata M, Tsuneki T and Ookawa M. *Stud Surf Sci Catal*, 108, 617–624 (1997).
555. Comparative study of hexamethyldisiloxane photofragmentation through multiphotonic and monophotonic processes—Quintella C.M, Desouza G.B and Mundim MS. *Proc SPIE Intl Soc Opt Engg*, 3271, 227–235 (1998)
556. Hydrosilylation of olefins catalyzed by activated titanocene prepared from the reduction of titanocene-dichloride with excess lithium—Lee S. J and Han B.H. *Main Grp Met Chem*, 21(6), 315–318 (1998).
557. Laser induced decomposition of silacyclobutane—Khachatryan L, Volnina E.A, Fajgar R and Pola *J Organomet Chem*, 566(1–2), 263–270 (1998).
558. Effect of ultra sound on some siloxanes—Quelhorst H and Binneweiss M, *Z. Anorg Allge Chemi*, 624(9), 1548–1550 (1998).
559. Rhodium mediated silylative cyclization of 1,6-Heptadienes—Muraoka T, Matsuda I Itoh K. *Tetrahedron Lett.*, 39(40), 7325–7328 (1998).

560. Platinum catalyzed silaborative coupling of 1,3-dienes to aldehydes: Regio and stereo selective alkylation with dienes through allylic platinum-intermediate—Sujinome M, Nakamura H, Matsuda T and Ito Y. *J Am Chem Soc*, 120(17), 4248–4249 (1998).
561. Thermal and Lewis acid catalyzed intramolecular ene-reaction of allenylsilanes—Weinreb S.M and Smith D.T. *Synthesis (Spec)*, 509–521 (1998).
562. Organosilanes in sulfur chemistry: Silicon mediated synthesis and reactivity of sulfur containing molecules—Degl' Innocenti A and Capruci A. *Sulfur Rep*, 20(3), 279–395 (1998).
563. Hydrosilylation catalysts and preparation of a silane by using them—Kuboto Y, Iijima T, Endo M and Kuboto T. Jpn Kokkai Tokkyo Koho JP10 85,605 (7 April, 1998).
564. Catalytic Si-C bond formation by nucleophilic at silicon by benzyl anions generated over KNH_2 loaded on alumina—Baba T, Yuasa H, Handa H and Ono Y. *Catal Lett*, 50(1–2), 83–85 (1998).
565. Catalytic transformation of oligocarbosilanes induced by $AlCl_3$—Chernyavskaya N.A, Aleksinskaya V.I, Zavin B.G and Belokon A.I. *Russ Chem Bull*, 46(12), 2147–2148 (1997).
566. Intramolecular radical chain hydrosilylation catalyzed by thiols: Cyclization of alkenyloxysilanes—Cai Y and Ruberts B.P. *J Chem Soc. Perkin Transac.*, 1(3), 467–476 (1998).
567. A straightforward route to polyenylsilanes by palladium or nickel catalyzed cross-coupling reaction—Babudri F, Farinola G.M, Fiandanase V, Mazzone L, and Naso F. *Tetrahedron*, 54(7), 1005–1094 (1998).
568. Highly planar silane $[(i\text{-pr})_3Si]_3SiH$ and silyl radical $[(i\text{-pr})_3Si]_3{\cdot}Si$—Kyushin S, Sakutai H and Matsumoto H. *Chem Lett.*, 2, 107–108 (1998).
569. A novel route to penta-coordinated organylsilanes and germanes—Gevorgyan V, Borisova L, Vyater A, Ryabova V and Lucivies E. *J Organomet Chem*, 548(2), 149–155 (1997).
570. Rhodium catalyzed insertion of carbenoids into beta C-H bond of silacycloalkenes: A facile and general approach to functionalised silacycloalkenes—Hatomaka Y, Watnabe M, Onozawa S, Tanaka M and Sakurai H. *J Org Chem*, 63(3), 422–423 (1998).
571. Platinum catalyzed regioselective silaboration of alkenes—Sujinome M, Nakamura H and Ito Y, *Angew Chem (Intl Ed)*, 36(22), 2516–2518 (1997).
572. Rhodium catalyzed intramolecular silyl formylation of alkenes—Leghton J.L and Chapman E, *J Am Chem Soc.*, 119(50), 12416–12417 (1997).
573. Catalytic dehydrocoupling of silane by a homogeneous rhodium complex with water—Shi M and Nicholas K.M. *J Chem Res Synop*, 11, 400–401 (1997).
574. Ruthenium catalyzed cross-metathesis of trisubstitued vinylsilanes with light alkanes—Foltynowich Z and Marcienec B, *Appl Organomet Chem* 11(8), 667–661 (1997).
575. *Process for preparation of E-allyl-silane using palladium catalyst and isocyanide*—Ito Y and Sigonome M. Jpn Kokkai Tokkyo Koho JP09 227,579 (2 September, 1997).
576. Dehydrogenative coupling of olefins with silicon compounds catalyzed by transition metal complexes—Marcienec B. *New J Chem*, 21(6–7), 815–824 (1997).
577. Synthesis of linear and cyclic unsaturated carbosiloxanes via catalytic condensation of divinyltetramethyldisiloxane—Marcienec B and Lewandowski M. *Tetrahedron Lett*, 38(21), 3777–3780 (1997).
578. Migration of trimethylsilyl group in 3-trimethylsilylthiopropyne—Medvedeva A.S and Novokshonov V.V. *Russ J Org Chem*, 34(9), 1355–1356 (1998).
579. Silylene reactions with nitrogen multiple bond; addition and rearrangement silicon compounds with strong intra-molecular steric interaction: Part 68—Weidenbruch M, Olthoffs S, Saak W and Marsmann H. *Eur J Inorg Chem*, 11, 1755–1758 (1998).
580. Stereoselective migration from silicon to carbon—Studer A, Bossart M and Steen H. *Tetrahedron Lett.*, 39(489), 8829–8832 (1998).

581. The Sila-Pummerer rearrangement of 3,3-dimethyl-3-silathiane-8-oxide—Kirpichenko S.V, Suslova E.N, Albanov A.I and Shainyan B.A. *Tetrahedron Lett*, 40(1), 185–188 (1999).
582. Silicon tethered ring closing metathesis reactions for self and cross coupling of alkenols—Hoye T.R and Promo M.A., *Tetrahedron Lett.*, 40(8), 1429–1432 (1999).
583. Fluoride triggered decomposition of m-siloxyphenyl substituted dioxetanes by intramolecular electron transfer—Nery A.L.P, Ropke S, Catalani L.H and Baoder W.J. *Tetrahedron Lett*, 40(13), 2243–2446 (1999).
584. Migration of aryl groups from silicon to carbon in alpha, beta-epoxysilanes: A new model for hypervalent-silicon study—Achmatowich B, Jankowski P, Wich J and Zarecki A. *J Organomet Chem*, 558(1–2), 227–230 (1998).
585. Irreversible isomerization of halofunctional tris (silyl) hydroxylamines by dyotropic rearrangement—Wolfgram R and Klingebiel U, *Z Anorg Allge Chemi*, 624(6), 1031–1034 (1998).
586. Thermal rearrangement of O-fluorosilyl-N, N-bis(trimethylsilyl) hydroxylamine—Wolfgram R and Klingebiel U. ibid, 624(6), 1035–1040 (1998).
587. Acid catalyzed rearrangement of alpha-hydroxycyclopropylsilane—Sakaguchi K, Fujita M and Ohfune Y. *Tetrahedron Lett*, 39(24), 4313–4316 (1998).
588. Molecular structure refinement of 4-bromobenzoyloxymethyl-trifluorosilane—Zelbst E.A, Ovechinnikov Y.E, Kashaev A.A, Struchkov Y.T and Yoronkov M.G. *J Struct. Chem*, 38(6), 993–995 (1998).
589. Mechanism of rearrangement and alkene addition/elimination reaction of $SiC_3H_9^+$—Ignatayev I.S and Sundius T, *Organometallics*, 17(13), 2819–2824 (1998).
590. Discovery of novel 1,2-migration of the trimethylsilyl group—Stuart A.M, Coe P.L and Moody D.J, *J Fluorine Chem.*, 88(2), 179–184 (1998).
591. Irreversible rearrangement of silyl groups between oxygen and nitrogen in tris (silyl) hydroxylamines—Wolfgramm R, Mueller T and Klingebiel U, *Organometallics*, 17(15), 3222–3226 (1998).
592. Unusual exchange of functional groups at the silicon and metal atoms—Levitskii M.M, Zavin B.G, Karpilovskaya N.V and Chernyvski A.I. *Russ Chem Bull*, 47(8). 1614–1616 (1998).
593. Effect of substituents on silicon upon the ring inversion of Silepins annulated with bicyclo-2,2,2-octene framework—Nishimaga T, Izukawa Y and Komatsu K. *Chem Lett.*, 3, 269–270 (1998).
594. Ligand exchange via coordinative Si-N bond cleavage and psuedorotation in neutral pentacoordinated silicon complex—Kalikhmann I and Kost D. *Organosilicon Chem III (Muench Silicontage) 3rd 1996*, pp 446–451 (pub. 1998).
595. Isomeric halosilylhydroxylamines: Preparation and thermal rearrangements—Wolfgramm R and Klingebiel U. ibid, pp 348–352 (1998).
596. Rearrangement of bis (hypersilyl) silylenes and related compounds: An unusual way to 3-membered rings—Klinkhammer K.W. ibid, pp 82–85 (1998).
597. Cage rearrangement of Silesesquioxanes—Rikowski E and Marsmann H.C. *Polyhedron*, 16(19), 3357–3361 (1997).
598. Thermal isomerization of a Tetrahydro-2,3-disilanaphthalene into a 1,2-disilacyclobutane—Krempner C, Kampe R and Octme H. ibid 52(7),815–818 (1997).
599. Migration of trimethylsilyl groups in enolates of silyl esters—Hudrlik P.F, Roberts R.R, Ma D and Hudrlik A.M. *Tetrahedron Lett*, 38 (23), 4029–4032 (1997).
600. Synthesis and reactivity studies of palladium complexes of bulky $CH(SiME_3)_2$ Group—Alias F, Bolderrain T, Carmona E, Graif C, Paneque M and Tiripicchio A, *Organomet Chem*, 577(2), 316–322 (1999).
601. Recent applications of organosilanes in the chemistry of sulfurated compds—Cappruci A, Deglibocenti A and Scafato P. *Rec Res Dev Synth Org Chem*, 1, 171–197 (1998).

602. Synthesis, characterization and structural elucidation of triorganosilicon (IV) complexes—Sharma R.K, Singh Y and Rai A.K. *Phos, Sulf, Silicon Rel. Elem.* 142, 249–257 (1998).
603. Oxidation of alpha-hydroxysilanes by leadtetraacetate—Paredes M.D and Alonso R. *Tetrahedron Lett,* 40(20), 3973–3976 (1999).
604. Enantioselective reduction of alpha, beta-unsaturated acylsilane by chiral lithium-amide—Takeda K, Omnishi Y and Koizumi T. *Org Lett,* 1(2), 237–239 (1999).
605. Reaction of lithium-silenolates with acylhalide for synthesis of di and tetra-acyl silane—Oshita J, Takunega Y, Sakurai H and Kunai A. *J Am Chem Soc* 121(25), 6080–6081 (1999).
606. Synthesis and thermolysis of 3-substituted 3-trimethyl-silylcyclopropene—Arrowood T.L and Kass S.R., *Tetrahedron,* 55(22), 6739–6748 (1999).
607. Synthesis of some mono- and diethyl-silane—Medvedeva A.S, Yazovtsev I.A, Demina M.M, Lyashenko G.S, Kozyreva O.B and Voronkov M.G., *Russ J Org Chem,* 34(9), 1263–1265 (1998).
608. Synthesis and certain transformation of silicon containing diacetylenicdiols—Novokshonov V.V, Medvedeva A.S, Demina M.M, Safronova L.P and Voronkov MG. ibid , 34(10), 1426–1431 (1998).
609. Effect of counteraction inclusion by Cryptand upon stabilization of potassium organofluorosilicates—Yamaguchi S, Akiyama S and Tamao K. *Organometallics,* 18(15) 2851–2854 (1999).
610. Friedel-Craft acylation of 2-trimethylsilylnorbornene effect of acyl group on the position of attack—Nagendrappa G and Begum N.S. *Tetrahedron,* 55(25), 7923–7934 (1999).
611. Reaction of chlorinated vinylsilane with hydrochloricacid—Lakhtin G, Ryabkov V.L, Kisin A.V and Nosova V.M. *Russ Chem Bull,* 48(2), 375–378 (1999).
612. Freeradical cyclization of trienes with this(trimethylsilyl) silane—Gomez F.J, Jaramillo G.L. and Hudlicky T. *Synth Commun,* 29(16), 2795–2806 (1999).
613. Remarkable enhancement of catalyst activity of Trialkylsilylsulfonates on the Mukaiyama reaction: A new approach using bulky organo-aluminium cocatalyst-Oishi M, Aratake S and Yamamato H. *J Am Chem Soc,* 120(32), 8271–8272 (1998).
614. *Process for preparation of 3-Glycidyloxypropyltrialkoxy-silanes*—Bade S, Monckiewicz J and Schon U. Eur pat Appl EP 934,947 (11 August, 1999).
615. *Preparation of 1,3-bis (3-aminopropyl) tetramethylsiloxane*—Khubota T and Endo M. Jpn Kokkai Tokkyo Koho JP11 209,385 (3 August, 1999).
616. Disproportionation of Triethylsilane over CaY-zeolite—Suzuki E, Nomoto Y, Okamoto M and Ono Y. *Catal Lett.,* 54(4), 223–225 (1998).
617. Modern organic silicon chemistry—Sakurai H. *Gendai Kagaku,* 331, 32–40 (1998).
618. Chemistry of organosilicon compounds (350): Highly selective monoand poly-allylation of polychlorosilane and polychlorodisilane—Sanji T, Iwata M. Watnabe M, Hoshi T and Sakurai H. *Organometallics,* 17(23), 5068–5071(1998).
619. Synthesis and structure of (4+2)-coordinated tetraorganosilicon compound—Mehring M, Jurkschat K and Schurman M. *Main Grp Chem,* 21(10), 635–641 (1998).
620. Reaction of Octamethylcyclotetrasiloxane with aluminium, galium and silicon iodides—Voronkov M.G and Dubinskaya E.I, Russ *J Gen Chem,* 68(4), 654–655 (1998).
621. Synthesis of a enantiometrically propargyl-silane—Fleming I and Mwamaki J.M, *J Chem Soc. Perkin Transac,* 1(7), 1237–1248 (1998).
622. Preparation of amino substituted cyclic-vinylsilane by Wurtz-Fittig reaction—Hariprasad S and Nagendrappa G. *Ind J Chem (Sect B),* 36B (11), 1016–1019 (1997).
623. *Preparation of vinyl-disiloxane*—Zhum V.I, Zhum A.B, Polivanov A.N and Chernysev E.A. Russ RU 2, 100,335 (27 December, 1997).
624. *Preparation of Methylchlorosilanes*—Levin Y.L and Grinberg E.E. Russ RU 2, 100,362 (27 December, 1997).

625. *Preparation of w, w′-dihydroxydimethylsiloxanes*—Shaptin A.S, Zhingalin G.Y, Pashintseva I.M, Ershov O.L, Shiminova N.I, Fedotov E,V, Polivanov A.N and Chernyshev E.A. Russ RU 2, 100,361 (27 December, 1997).
626. A zwitterionic penta-coordinated silicon-compound with an SiO_2N_2C framework—Pfrommer B and Tacke R. *Eur J Inorg Chem*, 3, 415–418 (1998).
627. Aryl containing Octasilsesquioxane (Vinylbenzene derivative): Perbenzyloctasilsesquioxane—Lavrentiev V.I. *Russ Chem Bull*, 67(2), 239–244 (1997).
628. Synthesis of spirocyclic aminosilanes withdrawing substituents at nitorgen—Wollenweber M, Kesse R and Stoeckii E.H. *Z Naturforsch B, Chem Sci*, 53 (2), 145–148 (1998).
629. *Method for preparing bis-triphenylsilanechromate*—Dai L, Duan Q and Ji H. Faming Zhuanli Shenquing Gongkai Shomingshu (China), CN 1, 143,082 (19 Feb, 1997).
630. Acyl-silanes—Page PCB, McKenzie M.J, Klair S.S, and Rosenthal S. *Chem Org Silicon Compd 2 (Pt II)*, 1599–1665 (1998).
631. Recent advances in hydrosilylation and related reactions—Ojima I, Li Z and Zhu J. ibid, 1687–1693 (1998).
632. Compounds of silicone (Part 124): Unsaturated compounds of silicon and its homolog. (Part 53): Sterically overloaded compounds of silicon. (Part 17): t-$(Bu_3Si)_4Si_4I_2$, a molecule containing an unsaturated Si_4 ring—Wiberg N, Auer H, Noth H, Knizek J and Polborn K. *Angew Chem. Intl Ed*. 32(20), 2869–2872 (1998).
633. Synthesis of Diethylsilane doubly labeled by tritium—Kochina T.A, Vrazhnov D.V, Nefedov V.D and Sinotova E.M. ibid, 68(6), 917–918 (1998).
634. Alpha elimination of silanones as a method of formation and decomposition of siloxanes—Voronkov M.G. ibid, 68(6), 901–906 (1998).
635. Synthetic applications of allylsilanes and vinylsilanes—Luh T.Y and Liu S, T. *Chem Org Silicon Compd 2 (Pt III)*, 1793–1868 (1998).
636. Cyclic-polychalcogenide compounds with silicon—Choi N and Ando E. ibid, 2 (Pt III), 1895–1928 (1998).
637. Cyclopentadienyl silicon compounds—Jutzi P. ibid, 2 (Pt III), 2129–2175 (1998).
638. Recent advances in the chemistry of cyclopolysilanes—Hugge E and Stugger H. ibid, 2, (Pt III), 2177–2216 (1998).
639. Inter- and intra-molecular oxidative addition of Si-H bond—Karsch R, Gilges H and Schubert U. *Organosilicon Chem III (Muench Silicontage) 3rd 1996*, pp 271–274 (Published 1998).
640. Novel synthetic approach to molybdenum-silicon compounds: Structure and reactivities—Jutzi P and Petri S.H.A. ibid, 275–280 (1998).
641. Silylhydrazine: Precursor for ring, hydrazones and pyrazolones—Drost C, Klingebiel U and Witte-Abel H. ibid, pp 358–363 (1998).
642. Some surprising chemistry of sterically hindered silanols—Lickiss P.D. ibid, pp 369–375 (1998).
643. Silanetriols: Preparation and their reactions—Murugavel R, Voigdt A. Walawalker M.G and Roesky H.W. ibid, pp 376–394 (1998).
644. Azomethine substituted organotrialkoxy-silanes and polysiloxanes—Mucha F and Roewer G. ibid, pp 400–402 (1998).
645. Metallo-silanols and metallo-siloxanes: Synthesis and reactivity of silicon transition metal complexes (Fe-complexes)—Reising J, Malisch W and Lankart R. ibid, 412–414 (1998).
646. Metallo-silanols and metallo-siloxanes (ferrio and tungsten silanols)—Malisch W, Reising J and Schneider M. ibid, pp 415–417 (1998).
647. Investigation of nucleophilic substitution at silicon: An unprecedented equilibrium between ionic and covalent chlorosilanes—Scharr D and Belzer J. ibid, pp 429–434 (1998).

648. Studies on the regio-selectivity of hydroformylation with alkenyl-alkoxysilanes—Wessel M and Grobe J. ibid., pp 538–542 (1998).
649. Selective hydrogenation of Methylchlorooligosilane—Herzig U and Roewer G. ibid., pp 312–316 (1998).
650. Polyhedral silicon compounds—Sekiguchi A and Nagasse S. ibid, 2 (Pt I), pp 119–152 (1998).
651. Structural chemistry of organosilicon compounds—Kaftory M, Kapon M and Botoshanski M. ibid, 2 (Part I), 181–265 (1998).
652. Synthesis and characterization of biophilic silicon organic compounds—Aisa AMA, Kraetzer D and Richter H. *Pharmazie*, 53(11), 751–758 (1998).
653. Chirality in bio-organosilicon chemistry—Tucke R and Wagner S.A. *Chem. Org Silicon Comp. 2(Part III)*, 2363–2400 (1998).
654. Activity and directive effect (on electrophillic substitution etc) of silicon—Bassindale A.R, Glynn S.J and Taylor P.G. ibid (Part I) 355–430 (1998).
655. Mechanism and structure in alcohol addition reaction of disilene and silenes—Sakurai H. ibid., 2 (Part I), pp 827–855 (1998).
656. Influence of water content on C. Cylindracea-lipase catalyzed bioconversion of organosilicon compounds—Qiu S, Yao R and Zhong M. *Guizhou Gongye Daxue Xuebao*, 27(4), 91–96 (1998).
657. *Preparation of 3-chloropropyltrichlorosilane*—Hu C. Faming Zhuanli Shenquing Gongkai Shinomigschu CN 1 123,283 (29 May, 1996).
658. *Branched higher alkylsilanes*—Arkles B.C. PCT Intl Appl WO 99 03,864 (28 January, 1999).
659. *Preparation of alkylhalosilanes*—Nakanishi T, Tsukioka K. Nakayama H and Satoh Y. Eur Pat Appl EP 893,448 (27 January, 1999).
660. *Matrix isolation studies of the reaction of silicon atoms*—Maier G, Reisenauer H.D and Egenolf H. ibid, pp 31–35 (1998).
661. *Matrix isolation of unusual silicon species: Oxidation of silenes and silylenes*—Sander W, Trommer M and Patyk A, ibid, pp 86–94 (1998).
662. *Silaspirocycles as precursors for a 2-silaallene*—Goetze B, Herrschaft B and Aumer N. ibid., pp 106–112 (1998).
663. *Dieno and enophilicity of sila, germa and stanna-ethenes*—Wiberg N and Wagner S. ibid., pp 117–119 (1998).
664. *Imminosilanes and silamides: Synthesis and reactions*—Niesmann J, Frenzel A and Klingebiel U. ibid., pp 120–125 (1998).
665. *1,2-ditertiarybutyl-tetrafluorosilane: a highly fluxional molecule*—Zink R, Hassler K, Mitzell N.W, Smart B.A and Reinkin D.W. ibid, pp 248–253 (1998).
666. *Unexpected reactivity of 1,2-bis (bromodiphenylmethyl) 1,1,2,2-tetramethyl disilane*—Pillong F, Schutt O and Strohmann C. ibid., 281–285 (1998).
667. *Chlorination of methylphenyloligosilanes: Products and reactions*—Notheir C, Brundler E and Thomas B. ibid, pp 307–311 (1998).
668. *On the reaction of* $(t\text{-}Bu_2SnO)_3$ *with organochlorosilane: Simple formation of* $(t\text{-}Bu_2O\text{-}SnO)_2(t\text{-}Bu_2SiO)$—Beckmann J, Jarkschart K and Shollmeyer D. ibid, pp 403–406 (1998).
669. Hexacoordinated carbon or tetracovalent silicon—Schiemenz G.P, Sciemenz B, Peterson S and Wolf C. *Chirality*, 10(1/2), 180–189 (1998).
670. *Preparation of alkoxysilane from silicon and alcohols*—Yamada Y, Harada M. Jpn Kokkai Tokkyo Koho JP11 21,288 (26 January, 1999).
671. Modern organosilicon chemistry—Sakurai H. *Gendai Kagaku*, 334, 56–63 (1999).
672. Synthesis and properties of isocyanato-disiloxane and their alkoxy substituted derivatives—Gunji T, Setogawa A, Asakura K and Abe Y. *Bull Chem Soc Jpn*. 71(12), 2967–2972 (1998).

673. Bimolecular reaction of tetrakis (trialkylsilyl) disilenes with various reagents—Iwamoto T, Sukurai H and Kira M. ibid, 71(12), 2741–2747 (1998).
674. Synthesis and characterization of the bulky terphenylsilanes and chlorosilane—Simon R.S, Haubrich S.T, Mork B.V and Niemeyer M. *Main Grp Chem*, 2(4), 275–283 (1998).
675. *Preparation of oxazolinyl-carbonyl-aminoalkylsilane*—Zaima H, Sato K. Jpn Kokkai Tokkyo Koho JP11 12,288 (19 January, 1999).
676. *Preparation of aminopropylsiloxnae*—Kobayashi H and Iwai A. ibid, 11 21, 289 (26 January, 1999).
677. The reaction of N-nitramines and their trimethylsilyl derivatives with N,N-bis (trimethylsilyloxy) enamines—Ioffe S.L, Makarenkova L.M, Strelenku Y.A, Bliznets I.V and Tartakovsky V.A. *Russ Chem Bull*, 47(10), 1989–1991 (1998).
678. Inclusion complexes of silicon-hydrofluoric acid transformation products with crown-ether—Simonov Y.A, Lipkowski J, Fenari M.S, Kravtsov V.C, Ganin E.V, Gelmboldt V.O and Ennan A.A. *Mol Recognit Inclusion (Proc Intl Symp. 9th, 1996)*, pp 503–506 (Pub 1998).
679. Reaction of silica with trichlorosulfophenylsilane—Belyakova L.A and Lyashenko D.Y. *Ukr Khim Zh* (Russ Ed), 63(11–12), 113–117 (1997).
680. Cis-hydroylation of cyclic vinylsilane using Cetyltrimethylammonium permanganate—Mallaya M.N and Nagendrappa G. *Synthesis*, 1, 37–39 (1999).
681. Stereospecific cationic (1, 2) silyl shift with retention of configuration at the migrating terminus—Sujinome M, Sujinome M, Takama A and Ito Y. *J Am Chem Soc*, 120(8), 1930–1931 (1998).
682. Silicon compounds with strong intramolecular steric interactions: Silylene and disilene addition to hexa-2,4-diyne: formation of a propynylsilirene and bicyclic compounds—Kirmaire L, Weidenbruch M, Marsmann H and Peter K. *Organometallics*, 17(6), 1237–1240 (1998).
683. Oxasilacyclopentanes as intermediates for silicon tethered ene cyclization—Robertson J, Middleton D.S, O'connor G and Sardharwala T. *Tetrahedron Lett*, 39(7), 669–672 (1998).
684. *Separation of hexamethylcyclotrisyloxane from cyclopolydimethylsiloxane mixtures*—Nomura T (Dow-Corning Torray Silicone Co, Japan). Jpn Kokkai Tokkyo Koho JP10 45,770 (17 February, 1998).
685. Nucleophillic addition of trimethylsilylamines to 2-halo-2-alkenals—Rulev A.Y, Kuznetsova T.A, Mokov A.S, Sherstyannikova I.V, Keiko N.A and Yoronkov M.G, *Russ J Org Chem*, 33(1), 26–28 (1997).
686. Planar hexasilylbenzene dianions with thermally accessible triplet state—Ebata K, Setaka W, Inoue W, Inoue T, Kabuto C, Kira M and Sakurai H. *J Am Chem Soc*, 120(6), 1335–1336 (1998).
687. Beta-donor bonds in compounds containing SiON fragments—Mitzel N.W and Losehand U. *Agnew Chem (Intl Ed)*, 36(24), 2807–2809 (1997).
688. Friedel-Crafts polyalkylation of alkylbenzene with dichloromethylvinyl-silane—Cho U.J, Lave V and Jung I.N, *J Organomet Chem*, 54(2), 237–245 (1997).
689. Cycloaddition of allylsilanes (Part 12): Regio and stereo selective transformation of silylbicycloalkenes—Knoelker H.J, Foitzik N, Gabler C and Graft R. *Synthesis*, 1, 145–151 (1999).
690. Pentacoordinated organo and hydrido cyclic silanes via sulfur donor action—Mercado R.M, Chandrashekharan A, Day R.O and Holmes R.R. *Organometallics*, 18(5) 905–914 (1999).
691. Ferrocinysiloxane chemistry: Synthesis and characterization of hexaferro-cenylcyclo-trisiloxane and tetraferrocenyl-disiloxanediol—Machachlam M.J, Zheng J, Lough A.J, Manners I, Mordas C, Lesuer R and Geiger W. *Organometallics*, 18(7), 1337–1345 (1999).

692. From di-iododimethylsilane to dimethylsilanone—Voronkov M.G, Tsyrendor zhieva I.P and Dubinskaya E.I. *Russ J Gen Chem*, 68(8), 1340–1341 (1998).
693. A facile synthesis of 2,6-bis(trialkylsilyl)-4H-pyran from 1,5-diacylsilane—Saleur D, *Tetrahedron Lett*, 40(10), 1885–1886 (1999).
694. Formation and structure of protonated tetrasilatetrahedranemonoxide (tert-$Bu_3Si)_4$ Si_4OH^+—Ichinobe M, Takahashi N. and Sukuguchi A. *Chem Lett*, 7, 553–554 (1999).
695. Synthesis and molecular structure of novel isopropyl substituted oligosilane—Tanaka R, Unno M. and Matsumoto H. *Chem Lett*, 7, 595–596 (1999).
696. Synthesis and reactivity of stannyl-oligosilanes (Part I): Stannyl oligosilane chains containing $SiMe_2$ moities—Uhlig F, Kayser C, Klassen R, Herman U, Brecker L, Schuermann M, Ruhland S.K and Englich U. Z *Naturforsch B. Chem Sci*, 54(2), 278–287 (1999).
697. Organosilicon compounds in cross-coupling reactions—Hiyama T. *Met Catal Cross Coupling Reaction*, 421–453 (1998).
698. Investigation of organosiloxane equilibrium reactions—Kapylov V.M and Kovyarin V.A. *Intl J Polym. Mater*, 38(1–2), 129–171 (1997).
699. Free and coordinated iminosilanes: synthesis and structure—Niesmann J, Klingebiel U, Schaefer M and Boese R. *Organometallics*, 17(5), 947–953 (1998).
700. Disproportionation of Triethoxysilane over KF/Al_2O_3 and heat treated hydrotalcite—Suzuki E, Nomoto Y, Okamoto M Ono Y. *Appl Catal*, 167(1), 7–10 (1998).
701. Nucleophillic fluorination of alkoxysilane with alkali metal salts of perfluorinated complex anions—Farroq O. *J Chem Soc (Perkin Transac)*, 1(4), 661–666 (1998).
702. Synthesis and structure of a series of Oligo-1,1-(2,3,4,5-tetramethylsilole)—Kanno K, Ichinohe M, Kabuto C and Kira M. *Chem Lett*, 1, 99–100 (1998).
703. Siloxanes as sources of Silanones—Chernyshev E.A, Krasnova T.L, Sargeev A.P and Abramova E.S. *Russ Chem Bull*, 46(9), 1586 –1589 (1997).
704. Preparation of (E)-1-alkenylthiosilanes by the reduction of silicon capture of 1 - alkenesulfenate anions—Schwan A.L and Refvik M.D *Synlett*, 1, 96–98 (1998).
705. *Tertiary alkylsilane*—Ikai K, Minami M and Shiozaki I. Eur Pat Appl EP 922,706 (16 January, 1999).
706. A simple approach to carbosiloxane dendrimers—Brunning K and Lang H. *J. Organomet* Chem, 575(1), 153–157 (1999).
707. *Preparation of gamma-methacrylo-oxapropyl-silanes*—Kuwayame J and Yada T. Jpn Kokkai Tokkyo Koho JP 11 100,387 (13 April 1999).
708. *Process for preparation of aminosilanes*—Ikai S, Fukunaga T and Fujimoto J. ibid, 11 106,389 (20 April, 1999).
709. *Preparation of a mixture of (3-aminopropyl) triethoxysilane and (2-aminopropyl) triethoxysilane*—Marciniec B, Gulinski J, Nowicka T and Oczkowich S. Pol PL 175,322 (31 December, 1990).
710. *Silicon (IV) reagents*—Hosomi A and Miura K. *Lewis Acid Reagents*, 159–168 (1999). Oxford University Press.
711. *Preparation of bis (polycyclicamino) dialkoxysilanes*—Fukunaga T and Aritomi T. Jpn Kokkai Tokkyo Koho JP11 130,780 (18 May, 1999).
712. *Preparation of bis (polycycliscamino) dialkoxysilanes*—Ikai S, Fukunaga T and Koizumi S. ibid, 11 130,785 (18 May, 1999).
713. Synthesis and crystal structure of addition compound with alkyldimethyl-bromosilane and N-trimethylsilylimidazole—Hensen K, Gebhardt F and Bolte M. Z *Naturforsch(B) Chem Sci*, 53(12), 1491–1496 (1997).
714. *Method of producing alkoxysilanes*—Weidner K and Bheml W (Wecker Chemie, GmbH) Ger Offen DE 19,629,760 (29 January, 1998).
715. *Method for preparation of silicon containing isocyanate compounds by the thermal decomposition of silicon containing carbamic-acid esters*—Uchida T, Okumara H, Shibata K and Sasaki S. Jpn Kokkai Tokkyo Koho JP10 01,486 (6 January, 1998).

716. *Preparation of aminopropyl-alkoxysilane derivatives*—Kubota T, Endo M and Hirahara T. ibid, 10 17,578 (20 January, 1998).
717. Nucleophilic fluorination of alkoxysilane with alkali-metal hexafluorophosphate—Farooq O, *J Fluorine Chem*, 86(2), 189–197 (1997).
718. Structural chemistry of Aziridino and Azetidino Silanes—Huber G, Jockish A and Schneidbaur H. *Eur J Inorg Chem*, 1, 107–112 (1998).
719. Cyclization and cyclopolymerization of silicon containing dienes: Functionalisation of carbosilane dendrimers—Krska S.W (M.I.T, USA); pp 6571 (1997); *Dissr Abstr Intl B*, 58(12), 6571 (1998).
720. Preparation of amino functionalised cyclotrisiloxanes via salt elimination reaction—Veith M and Rammo A. *Phos. Sulf. Silicon Rel. Elem.* 123, 75–87 (1997).
721. Preparation of Allylsilanes by olefination of carbonyl compounds using betasilylthioacetals—Takeda T, Watnabe M, Rahim A and Fujiwara T. *Tetrahedron Lett.*, 39(22), 3753–3756 (1998).
722. Rhodium catalyzed asymmetric hydrosilylation using new oxazoline ligands with potential charge transfer properties—Brunner H and Stoeriko R. *Eur J Inorg Chem*, 6, 763–788 (1998).
723. Reaction of silicon containing alpha beta-acetylinic aldimines with diazomethane—Khramchikhin A. V, Proshkin A.I, Picterskaya Y.L and Standnichuk M.D. *Russ J Gen Chem*, 67(11), 1816–1817 (1997).
724. Reaction of the Si-Cl bond with Trialkylorthoformates: Preparation of alkoxy substituted silanes—Herzog U, Schulze N, Trommer K and Roewer G. *J Organomet Chem*, 547(1), 133–139 (1997).
725. Generation and decomposition of penta-coordinated Spiro-bis (1,2-oxasiletanide)—Kawashima T, Nagamuma K and Okazaki R. *Organometallics*, 17(3), 367–372 (1998).
726. Synthesis and hydrolysis of hybridized silicon-alkoxides—Cao B, Tang Y, Zhu C and Zhang Z. *J Sol-Gel Technol*, 10(3), 247–253 (1997).
727. *Oligomeric mixture of condensed alkyl-alkoxysilanes*—Standke B, Edelmann R, Frings A, Hora M and Junkner P. Ger Offen DE 19,624,032 (18 December, 1997).
728. Highly branched dendritic macromolecules with core polyhedral Silesesquioxane functionalities—Hong B, Thomas TPS, Murfie H, J and Lebrum M.J, *Inorg Chem*, 36(27), 6146–6147 (1997).
729. Substituent effect on the silicon-carbon double bond: Arrhenius parameter for the reaction of 1,1-Diarylsilanes with alcohols and acetic-acid—Bradaric CJ and Leihgh W.J, *Can J Chem*, 75(10), 1393–1402 (1997).
730. Access to iminosilicates from novel triminosilanes: A short review—Rennekamp C, Wessel H and Roesky H.W. *Phos. Sulf. Silicon Rel Elem*, 124–125, 275–184 (1997).
731. Preparation and demethylation of Hexamethyldisilane—HuC.Y, Guo Z, Huang G.L and Pan Y.Y. *Hecheng Huaxue*, 6(1), 61–64 (1998).
732. Chemistry of the direct synthesis of Methylchlorosilane—Sun D.H, Bent B.E, Wright A.P and Naasz B.M *J Mol Catal(A): Chem*, 131(1–30), 169–183 (1998).
733. Hetero pi-system (Part 28): Reaction of silicon atom with acetylene and ethylene (generation and matrix identification spectroscopically of C_2H_2Si and C_2H_5Si isomers)—Maier G, Reisenmauer H.P and Egenolf H. *Eur J Org Chem*, 7, 1313–1317 (1998).
734. Synthesis of novel siloxane containing aromatic dicarboxylic acids—Osadechev A.Y and Skorokhodov S.S. ibid, 71(4), 699–701 (1998).
735. Hydroxylamino-silanes: Compounds with beta-donor-acceptor bonds—Loschand U and Mitzel N.W. *Inorg Chem*, 37(13), 3175–3182 (1998).
736. Heats of Lewis base complexation, deaggregation and stabilization by alphasilicon in a family of primary alkyl-lithiums—Unno M, Takada K and Matsumito H. *Chem Lett*, 6, 489–490 (1998).

737. Synthesis, structure and reaction of Tetrahydroxycyclotetrasiloxane—Unno M, Takada K and Matsumito H. *Chem Lett*, 6, 489–490 (1998).
738. Silicon-carbon and silicon-nitrogen multiply bonded compounds—Muller T, Ziche W and Auner N. *Chem Org Silicon Compd 2 (Pt. II)*, 857-1062 (1998).
739. Vinylsilanes in synthesis of 2-halo-1-cyclo-pentenyl alkyl/aryl ketones from 2-halo-1-trimethylsilylcyclo-pentenes—Prasad S and Nagendrappa G. Indian *J Chem (Sect B), Org Chem*, 36B (8) 691–694 (1997).
740. Dimethoxysilane formation by the disproportionation of triethoxysilane over heat treated $Ca(OH)_2$—Suzuki E, Nomoto Y, Okamoto M and Ono Y. *Catal Lett*, 48(1–2), 75–78 (1997).
741. Reaction of hydrosilanes with lithium: formation of silole anions from 1-methylsilole via carbodianion—Wakahara T and Ando W. *Chem Lett*, 11, 1179–1180 (1997).
742. Synthesis, derivatization and structure of the Silanetriol $C_6H_4(SiMe_3)$. Si $(OH)_3$—Jutzi P, Schneider M, Stammler H. G and Neumann B. *Organometallics*, 16(24), 5377–5380 (1997).
743. Synthesis of potential cyclopropenyl anion precursors: 3-methyl-3-trimethyl silylcyclopropene and its dibenzoyl derivative—Hans S and Kass S.R. *Tetrahedron Lett*, 38(43), 7503–7506 (1997).
744. Preparation of acyloxysilanes from hydrosilanes and carboxylic acids—Kizaki Y. ibid, 10 182,666 (7 July, 1998).
745. Gas phase reactivity of p-Me3Si-substituted 1,3-diphenylpropane towards charged electrophiles: intra and inter-annular hydrogen migraiton—Crestoni M.E. *Chem Eur J*, 4(6), 993–999 (1998).
746. Highly regioselective palladium catalyzed internal arylation of allyltrimethyl-silane with aryl-triflates—Olofsson K, Larhed M and Nallberg A. *J. Org Chem*, 63(15), 5076–5079 (1998).
747. Efficient synthesis of (carboranylmethyl)disiloxane from carboranyl-copper and chloromethylsiloxane—Izmailov B.A, Neddkin V.I and Gerr I.S. *Russ Chem Bull*, 47(4), 687–690 (1998).
748. Synthesis and structure of silicon compounds intramolecularly cordinated by hydrazino groups—Belzner J.S. Dirk H.E, Kneisei B.O and Nottemeyer M. *Tetrahedron*, 54(29), 8481–8500 (1998).
749. Synthesis of silicon heterocycles via gas phase cycloaddition of amino-methylsilylene—Heinicke J and Meinel S. *J Organomet Chem*, 561(1–2), 121–129 (1998).
750. Induction of helical chirality in linear oligosilanes by terminal chiral substiuents—Obata K, Kabuto C and Kira M, *J Am Chem Soc*, 119(46), 11345–11346 (1997).
751. Selective synthesis of methylvinyl-dichlorosilane by hydrosilylation—Lan Z, Li F, Liu W and Zhan X, *Huaxue Tongbao*, 10, 39–42 (1997).
752. Unexpected reaction of benzoates with chlorovinylsilanes in the presence of magnesium—Tongco E.C, Wang Q and Prakash GKS. *Synthesis*, 9, 1081–1084 (1997).
753. Synthesis of thermally stable formaldehyde N-silylhydrazones—Klingebiel U, Kinpping K and Wettable H. *Z naturforsch B: Chem Sci*, 52(9), 1049–1050 (1997).
754. *Preparation of alkoxy (organo) silanes*—Klokov B.A, Trivanov V.D, Isaeva O.V, Savina T.M, Minkova N.I and Kulinskij A.I. Russ RU 2,079,502 (20 May, 1997).
755. *Preparation of hexamethyl-disilazane by ammonolysis of trimethyl-chlorosilane*—Matveev L.G, Natejkina L.I, Kozlov A.I, Polkin N.M, Alksyndrichev M.A and Shkuro V.G. ibid, 2,079,500 (20 May, 1997).
756. Asymmetric oligocyclic (dialkylsilethynes)—Voronkov M.G, Zhilitskaya L.V, Yarosh O.G, Lovanov A.I and Klyba L.V. *Russ J Gen Chem*, 67(12), 1893–1895 (1997).
757. Reaction studies on hypervalent silicon-hydride compounds—Weinmann M, Walter O, Huttner G and Lang H. *J Organomet Chem*, 561(1–2), 131–141 (1998).

758. Synthesis of alpha-substituted alpha, beta-unsaturated acylsilane—Tinus M.A and Hu H. *Tetrahedron Lett*, 39(33), 5937–5940 (1998).
759. Penta-coordinated chlorosilanes: Reaction chemistry, structure, bonding—Weinmann M, Gehrig A, Schiemenz B, Huttner G, Nuber B, Rheinwald G and Lang H. *J Organomet Chem*, 563(1–2), 61–72 (1998).
760. *Method for making organosilicon carbonyl compounds*—Graivers D, Khien AQ and Nguyen B.T. Eur Pat Appl EP 863, 146 (9 September, 1998).
761. *Preparation of hydrocarboxylated silicon compounds*—Inoue K, Iwata K, Ishikawa J, Fujikake S and Ito M. Jpn Kokkai Tokkyo Koho JP10 218,882 (18 August, 1998).
762. *Preparation of hydrocarboxylated silicon compounds*—Matsumoto M and Murakami S. ibid, JP10 218,885 (18 August, 1998).
763. Preparation of dimethoxymethylsilane from methylorthoesters and polyhydrogen-siloxanes—Ogowa N and Asai Y. ibid, 10,251,275 (21 September, 1998).
764. Preparation of Dimethoxymethylsilane from methylorthoesters and polyhydrogen-siloxanes—Ogawa N and Asai Y. ibid, 10,251,276 (22 September, 1998).
765. Synthesis of 1, 3-bis(silyl)cyclodisilazane—Jaschke B, Herbst I.R, Klingebiel U, Nengebauer P and Pape T. *J Chem Soc. Dalton Transac*, 18, 2953–2954 (1998).
766. *The chemistry of organic-silicon*—Rappaport Z and Apeloig Y, pp 2830, Wiley Chicester, U.K Publ. (1998).
767. Oligo (alkynyl) silanes: Templets for organometallic polymers—Kuhnen T, Stradiotto M, Ruffolo R, Ulbrich D, McGlinchey MJ and Brook MA. *Organometallics* 16(23), 5048–5057 (1997).
768. Olefinic aldoh reaction: addition of zincated hydrazone to vinylsilane—Nakamura E and Kubota K. *Tetrahedron Lett*, 38(40), 7099–7102 (1997).
769. Base promoted preparation of alkenyl-silanols from allylsilanes—Akiyama T and Imazeki S. *Chem Lett*, 10, 1077–1078 (1997).
770. Synthesis of aryl-silanes via palladium catalyzed silylation of arylhalides with hydrosilane—Murata M, Suzuki K, Watnabe S and Masuda Y. *J Org. Chem*, 62(24), 8569–8571 (1997).
771. *Preparation of halogenated alkyl-oligo-organosiloxane*—Zavin B.G, Kireev V.V, Astrina V.I and Pilipkova A.Y. Russ RU 2,078,766 (10 May, 1997).
772. *Preparation of phenylethoxysilanes*—Klokov B.A, Sidorov S.A. Tivanov V. D. Khe L.N, Isaeva O.V, Avdonin V.M and Kozlov V.P. ibid, 2,080,323 (27 May, 1997).
773. Stereoselective epoxidation of acylic allylic ethers using ketone-oxone system—Kurihara M, Ishi K, Kasschare Y, Kameda M, Pathak A.K and Miyata N. *Chem Lett*, 10, 1015–1016 (1997).
774. *Preparation of 4-trimethylsiloxy-3-penten-2-one*—Yahata T, Endo M and Kubata T. Jpn Kokkai Tokkyo Koho JP10 279,585 (20 October, 1998).
775. *Preparation of cyclic silicon compounds having functional group*—Hatanaka Y, Onozawa T and Tanaka M. ibid, JP10 251,271 (22 September, 1998).
776. Diel-Alder reaction of C-phenylated siloles with 1, 4-epoxy-1,4-dihydronaphthalene—Kirin S., Vikic T.D, Mestovic E, Kaitner B and Eckert M?M. *J Organomet Chem*, 566(1–2), 85–91 (1998).
777. Generation and capture of methyl(vinyl) silanone—Basenko S.V, Gebel I.A, Voronkov M.G, Klypa L.V and Mirskov R.G, *Russ Chem Bull*, 47(8), 1571–1573 (1998).
778. 2-trimethylsilylethane-sulfonylchloride from vinyltrimethylsilane—Heinl zelman G.R and Boeckman R.K, *Org Synth*, 75, 161–169 (1998).
779. Synthesis of 1,3-dichloro-1,2,3,3-tetramethyl-1-vinyl disilazane and its reaction with primary-amines—Ramkrishna T.V.V. and Elias A.J. *Phos. Sulf. Silicon Rel. Elem.* 130, 211–216 (1997).
780. Synthesis of some di and tri-cyclic silaalkanes—Teng Z, Boss C and Keese R. *Tetrahedron*, 53(38), 12979–12990 (1997).

781. Synthesis of flash vacuum thermolysis of silanethiones—Lefevre V and Ripoll J.L, *Phos. Sulf. Silicon Rel Elem*, 120–121, 371–372 (1997).
782. Application of chiral thiazolidine ligands to asymmetric hydrosilation—Li H, Yao J and He B. *Sci China: Ser B, Chem*, 40(5), 485–490 (1997).
783. Preparation and characterization of Octasilsesquioxane cage monomer—Harrison P.G and Hall C. *Main Grp Met Chem*, 20(8), 515–529 (1997).
784. Synthesis of silirenes by palladium catalyzed transfer of silylenes to alkynes—Palmer WS and Woerpel PA, *Organometallics*, 16(22), 4824–4827 (1997).
785. Stereospecific synthesis of tetrasubstituted Z-enolsilylethers by a three component coupling process—Corey E. J, Lin S and Luo G. *Tetrahedron Lett*, 38(33), 5571–5774 (1997).
786. *Process for preparation of organosilicon-disulfide compounds*—Cohen M.P and Parker D.K. Eur Pat Appl EP 785,207 (23 July, 1997).
787. *Preparation of fluoroenyl substituted organosilanes by Friedel-Crafts alkylation of biphenyls*—Cheng I, Yoo B and Hahn J. Jpn Kokkai Tokkyo Koho JP10 265,482 (6 October, 1998).
788. Preparation of halo (pentafluorophenyl) silanes, $(C_6F_5)_nSiX_{4-n}$ (X= F, Cl, Br; n = 2,3) from pentafluorophenyl(phenyl) silanes, $(C_6F_5)_nSiPh_{4-n}$— Frohn H. J, Lewin A and Bardin V.V, *J Organomet Chem*, 568(1–2), 233–240 (1998).
789. Synthesis and crystal structure of a nanometer scale dendritic polysilane—Lambert J.B and Wu H. *Organometallics*, 17(22), 4904–4909 (1998).
790. Synthesis and mesoporphic properties of 1, 10-dialkylpermethyl-decasilane—Yatabe T, Kanaiwa T, Sakurai H, Okumoto H and Tanabe Y. *Chem Lett*, 4, 345–346 (1998).
791. *Preparation of organosilicon compounds*—Takei M, Sumi A and Kimura K. Ger Offen DE 19, 703,695 (7 August, 1997).
792. Synthesis and structural characterization of functional bicyclic intramolecularly coordinated aminoaryl-silanes in series of dibenzol (1,5) aszasilocines—Carre F.H, Corriu RJP, Lannean G.F, Marle P, Soulariol F and Yao J, *Organometallics* 16(18), 3878–3888 (1997).
793. (Difluorophosphoryl) cyclosilazanes: Synthesis and crystal structure—Klingebiel U, Noltenmeyer M and Rakebrandt H. Z *Naturforsch B: Chem Sci*, 52(7), 775–777 (1997).
794. Study of the transmetalation of the silicon derivatives of O-carboranes on treatment with Bu_4Li—Zhakharkin L.I, 0l'Stevskaya V.A, Guseva V.V and Shemayakin N.F. *Russ Chem Bull*, 47(3), 475–477 (1998).
795. Transesterification of ethoxysilanes with butylacetate in the presence of alkoxy derivatives of titanium—Khomina T.G, Suvorov A.L, Soldatova E.E and Kozlov A.V. *Russ J Gen Chem*, 67(5), 729–732 (1997).
796. Gas phase reaction of Hexachlorodisilane with acetone—Chernyshev E.A, Komalenkova N.G, Kapitova I.V. Bykovchenko V. G, Kromykh N.N and Bochkarev V.N. ibid, 67(5), 7530755 (1997).
797. Improved method of synthesis of two silicon containing aromatic dianhydrides—Nuang W and Pan G. *Huaxue Shiji (China)*, 20(1), 54–55 (1998).
798. Per(tert-butoxy) and per(trimethylsiloxy) cyclosiloxanes—Basenko S.V and Voronkov M.G. *Russ J Gen Chem*, 67(7), 1029–1031 (1997).
799. Reaction of Hexachlorodisilane with methallylchloride in gas phase: synthesis of Trichloro (2-methyl-2-propenyl)silane—Chernysev E.A, Komalenkova N.G, Kapitova I.A, Bykovchenko V.G, Kromykh N.N and Bochkarev V.N. ibid, 67(7), 1040–1042 (1997).
800. *Process for the preparation of organosilicon-disulfide compounds*—Cohen MP and Wideman L.G (Goodyear Tyre and Rubber Co, USA). Eur Pat EP 785, 206 (22 Jan, 1996).
801. *Preparation of tris (Organosiloxy) antimony and their hydrolyzate SbO_3*—Hassengawa M and Matsumoto H. Jpn Kokkai Tokkyo Koho JP09 176,176 (8 July, 1997).

802. Aminosilanes and their application in organic synthesis—Rong G, Ma R and Long L. *Haxue Shiji*, 19(3), 147–149 (1997).
803. *Preparation vinylether substituted alkoxysilanes*—Kroener H, Friedrich H, Heider M and Gerst M (BASF-AG). Ger Offen DE 19,649,998 (19 June, 1997).
804. *Preparation of 3-chloropropylsilane*—Kiyomori A, Endo M, Kubota T, Kubota Y and Takenchi M. Jpn Kokkai Tokkyo Koho JP09 157276 (17 June, 1997).
805. *Preparation of alkoxysilanes from chlorosilanes and alcohols*—Kubota T, Endo M and Numanami K (Shin-Etsu Chemical, Japan), ibid, 09 157,277 (17 June, 1997).
806. *Composition for preparation of organo-monoalkoxysilanes*—Tsuji R. ibid 09, 157,278 (17 June, 1997).
807. *Preparation of organo-monoalkoxysilanes*—Tsuji R (Kanegafuchi Chemical Industries, Japan). ibid, 09, 157,279 (17 June, 1997).
808. Synthesis and properties of chlorinated and brominated aryltetrasilanes and aryltetrasilanes—Hassler K and Koell W. *J Organomet Chem*, 538(1–2), 135–143 (1997).
809. Bispropargylic bis-silanes as precursors of bis-vinylideneoxanes and oxepanes—Auburt P and Pornet *J. Organomet Chem*, 538(1–2), 211–21 (1997).
810. *Preparation and use of Dialkoxydiformoxysilane*—Friedrich H, Letner B, Mrong N and Schmidt R (BASF-AG). Ger Offen DE 19,545,093 (5 June, 1997).
811. Silicon-cobalt cluster with an unusually short cross-ring Si-Si distance—Borug S, Borug B, Carre F and Corriu RJP. *Organometallics*, 16(14), 3097–3099 (1997).
812. *Process for preparation of Si-H bond containing chloroorganosilanes*—Geisberger G and Lindner T. Eur Pat Appl EP 776,698 (4 June, 1997).
813. *Preparation of Cyclosiloxane derivatives by catalytic addition reaction*—Shin-Etsu Chemical, Japan, Jpn Kokkai Tokkyo Koho JP09 169,783 (30 June, 1997).
814. Synthesis and structural characterization of a remarkably stable, anionic incompletely condensed silsequioxane framework—Fehr F.G, Philps S.H and Ziller J.W. *Chem Commun (Cambridge)*, 9, 829–830 (1997).
815. Silicon controlled C-C bond formation and cyclization between carbonyl compds, and allyltrimethylsilane—Hwu JR, Shiao SS and Hakimelahi GH, *Appl Organomet Chem*, 11(5), 381–391 (1997).
816. Synthesis of Dialkoxydimethylsilanes and 2,2-Dimethyl-1,3-dioxa-1-silacyclo-alkenes—Lin J.M, Zhou A.M and Zhang A.H. *Synth Commun*, 27(14), 2527–2532 (1997).
817. *Method for processing chlorosilane containing mixtures*—Klementev I.Y, Gorshkov V.V and Petrovine N.M. Russ RU 2,068,848 (10 November, 1996).
818. *Synthesis of quinolone antibiotics by DIVERSOMER™ Technology: Innovation and perspective in solid phase synthesis*—Ed by Epton R, SPCC Ltd, Birmingham, AL, USA (1995). Diversomer Tech. invented by Furuka et. al.
819. *Cominatorial Chemistry: Synthesis and application*—Ed by Wilson S.R and Czarnik A.W, John-Wiley & Sons Inc, N.Y. (1997).
820. *Organosilicon compounds and their use in combinatorial chemistry* —Hone N.D and Bexter A.D. PCT Intl Appl WO 98 05, 671 (12 February, 1998).
821. Active Metal Powders—Riecke R.D. *Crit. Rev. Surf. Chem*, 1, 131–166 (1991).
822. Kenrick M.L and Kanner B (Union-Carbide), U.S 4, 593,114 (1985).
823. Wright J.R, Bolt R.O, Goldschmidt A and Abot A.D. *J Am Chem Soc*, 80, 1733 (1958).
824. Sumrell G and Ham G.E. ibid, 78, 5573 (1956).
825. Sellers J.E and Davis J.L (General Electric) U.S. Pat 2449821 (8 December, 1945).
826. Nitzsche S (Wecker Chemie). German Pat 906455 (23 February, 1949).
827. Weise J and Weltz H (Farbenfabriken, Bayer) Ger Pat 859164 (27 Feb, 1951).
828. Barton T Igadi-Maghsoodi S. U.S Pat 4, 940,967 (1990).
829. Rochow E.G (General-Electric). U.S. Pat 2380995 (26 September, 1941).
830. Dow-Chemical, Brit patent 609507 (U.S. Prior 15 January, 1945).
831. Dandt W,H (Dow-Corning). U.S. Pat 2672475 (1 December, 1953).

832. Rochow E.G (General-Electric). U.S. Pat 2459539 (22 May, 1947).
833. Dilthey W and Ednardoff F. *Ber Seutsche Chem Ges*, 37, 1139–1142 (1904).
834. Kipping F.S, *Proc Chem Soc (London)*, 20 15–16 (1904).
835. Gilman H and Zuech A. *J Am Chem Soc*, 79, 4560–4561 (1957).
836. Price E.P. *J AM Chem Soc*, 69, 2600 (1947).
837. Barry A.J (Dow-Chemical). Brit Pat. 618403 (U.S. Prior, 4 October, 1945).
838. Dow-Corning, Brit Patent 647940 (U.S. Prior, 2 October, 1944).
839. Emeleus H. J, and Robinson S.R. *J. Chem Soc (London)*, 1592 (1957).
840. Lewis R.N. *J Am Chem Soc*, 69, 717 (1947).
841. Rüst J.B. and Mckenzie C.A. (Montclair Res). U.S Pat 2426122 (17 October, 1944).
842. Gilman H and Zuech E.A. *J Am Chem Soc*, 79, 4560 (1957).
843. Rochoo E.G (General-Electric). U.S Pat 2459539 (22 May, 1947).
844. Tyler L. J, Sommer L. H and Whitemore F.C. *J Am Chem Soc*, 70, 2876 (1948).
845. Nebergall W.H and Johnson O.H. ibid, 71, 4022 (1949).
846. Tiganik L and Aktiebolag U. U.S. Pat 2521267 (Swed Prior, 8 September, 1947).
847. Gilman H and Massie S.P. *J Am Chem Soc*, 68, 1128 (1946).
848. Meals R.N. ibid, 68, 1880 (1946).
849. Nebergall W.H. ibid, 72, 4702 (1950).
850. Midland Silicones, Brit Patent 864848 (U.S. Prior, 19 January, 1959).
851. Geyer A.M, Haszeldine R.N, Leedhan K and Marklow R.I. *J Chem Soc (London)*, 4472 (1957).
852. Tarrant P, Dyckes G.W, Dunmite R and Butler G.B. *J Am Chem Soc*, 79, 6536 (1957).
853. Haszeldine R.N and Marklow R.J. *J Chem Soc (London)*, 962 (1956).
854. Wagner G.H (Union Carbide). U.S. Pat 263777 (17 September, 1949).
855. Noll W (Farbenfabriken Bayer). Ger Patent 825087 (15 August, 1949).
856. Shoshestvenskaya E.M, *Russ J Gen Chem*, 26, 247 (1956).
857. Wagner G.H (Union-Carbide): U.S Patent 37738 (1953).
858. *Comprehensive Handbook on Hydrosilylation*—Ed by Marciniec B (*Adv Organo Met Chem*—Speir J.L), pp 497–447. Pergamon Press N.Y (1992).
859. Nayes S.E (General Electric). U.S. Pat 4395563 (1981).
860. Ganeberg A, Vadevelde, Industrie Chemique Belg No. 6,591 (1980).
861. Schinabeck A *et al.* (Wacker-Chemie). U.S. 4298753 (1980).
862. Ayen R.J and Burk J.H. Mater Res Soc Symp (Proc), 73, 801–808 (1986).
863. Goodwin G.B and Kenney M.F. *Inorg Chem*, 29, 1216–1222 (1990).
864. Weaver D.G and O' Connors R. J. *Ind Eng Chem*, 50,132 (1958).
865. Bazant V, *Pure Appl Chem*, 13,313 (1996).
866. General Electric. Belg Pat 557059 (U.S. Prior 30 April 1956).
867. Speier J.L and Shorr L.M (Dow-Corning). U.S Pat 2811542 (5 August, 1953).
868. Noll J.E Speier J.L and Daubert B.F. *J Am Chem Soc*, 73, 3867 (1951).
869. Mitchell T.D (General-Electric). U.S. Pat 4602094 (1985).
870. Burkhardt J, Strekkel W and Boeck A (Wacker-Chemie) EPO 258640 (1986).
871. Burkhardt J (Wacker-Chemie) EPO 208285 (1985).
872. Wohlfahrt E, Nitzsche S and Hechtl W. U.S. Pat 3839388 (1972).
873. Rochow E.G (General-Electric). U.S. Pat 2258220 (27 April, 1940).
874. An Introduction to the Chemistry of Silicones—Rochow E.G, John-Wiley & Sons, N.Y, 2nd Ed (1951).
875. *The Nature of Chemical Bond*—Pauling L, Cornell University Press, Ithaca N.Y (1960).
876. Simmler W and Kauezor H.W (Farbenfabriken Bayer). Fr Patent 1326879 (1st April, 1963).
877. *Encyclopedia of Polymer Science and Technol (Vol. 12)*—Bikales N.M (Ed), Interscience Publishers (a Divn of John-Wiley & sons), N.Y (1970).

878. The status of crystalline silicon module—Margadona D and Ferraza F. *Renewable Energy*, 15(1–4), 83–88 (1998).
879. Hunter D, *Chemical Weekly*, pp 24–25 (19 February, 1992).
880. Silicon to Silicones: A partnership for growth—MayJames B. *Silicon Chem Indus IV Conf (Proc)*, pp 9–12 (1998).
881. Managing a technical revolution: the switch from Trichlorosilane to Trimethoxysilane—Ritscher J.S. *Silicon Chem Indus IV Conf (Proc)*. pp 265–273 (1998).
882. *The chemistry of organic silicon compounds*—Ed by Patai S and Rapaport Z, Wiley-Interscience, Chichester, UK (1989), pp 24 by Corey J.Y.
883. Photovoltaic technology expansion in future market (a review)—Tokoka H and Sawai H (Sharp Corpn, Technology Center, Tenri, Jpn). *Shapu Giho*, 70, 15–18 (1998).
884. Real spread of photovoltaic residential system—Sawai H. ibid,70. 44–48 (1998).
885. Photovoltaic system for residential use—Watnabe M and Haruyama T. ibid, 70, 54–58 (1998).
886. Data on fluorosilicone-Rubber available from Dow Corning Corpn, Midland, Michigan, or from General Electric Co, Waterford, N.Y (USA).
887. Tacke R *et al.*, *J Organomet Chem*, 417, 339–353 (1991).
888. Environmental aspects of silicon industry—Tveit H, *Silicon Chem Indus IV Conf (Proc)* (Ed by Oeye H.A), pp 365–376 (1998).